AF531618

About the Authors

Dr. S. Prabakaran has done his B.Sc. (Agri.) from TNAU sub-campus at Trichy. He studied his M.Sc. and Ph.D. from Indian Agricultural Research Institute (IARI), New Delhi. His area of specialisation was Plant Genetic Resources. In 2015, he received ICAR-JRF 22nd rank in plant sciences and secured merit fellowship. In 2017, he secured 1st rank IARI entrance exam and secured IARI merit fellowship for Ph.D. program. He also qualified Agricultural Scientist Recruitment Board (ASRB) – National Eligibility Test (NET) in 2017. He has qualified UGC-SRF and lectureship in Environmental sciences in 2019. Currently, he is teaching agriculture optional o civil service aspirants at Chennai.

A. Krishnamoorthi I did my Bsc agriculture (2012-16) from VIA Agriculutre college Affiliated to TNAU and I did my Msc Agriculture in Plant geneticr esource (2108-20) from TNAU and I am written 10 book chapters 2 research article and 10 review articles I cleared my ICAR SRF Examination with Rank 4 in the Discipline of Economic Botany and Plant Genetic Resources. Now I am pursuing my PhD at the ICAR-ational Bureau of Plant Genetic Resourrces IARI Pusa Campus New Delhi.

Glimpses of PGR

Glimpses of PGR

S. Prabakaran
A. Krishnamoorthi

1168, Sector 13, Urban Estate
Karnal-132 001, Haryana
Tel: 91-84470 75807, 18440 41168
Email: contentvibesppa@gmail.com
www.contentvibes.in

Print ISBN: 978-93-6134-178-6

ebook ISBN: 978-93-6134-790-0

Preface

Welcome to the first volume of "Glimpses of PGR," a collection that encapsulates the essence and evolution of the PGR community. This anthology brings together a diverse range of works that highlight the creativity, insights, and unique perspectives of its members.

Over the years, the PGR community has blossomed into a vibrant hub of intellectual exchange and artistic expression. Our members, hailing from various backgrounds and disciplines, contribute to a rich tapestry of thoughts and ideas. This volume is a testament to their dedication and talent.

In compiling this volume, our aim was to capture the spirit of PGR—its commitment to fostering innovation, its celebration of diverse voices, and its unwavering pursuit of excellence. Each piece in this collection has been carefully selected to reflect these core values. From thought-provoking essays to imaginative short stories, from poignant poems to ground-breaking research, the works presented here offer a glimpse into the dynamic world of PGR.

As you journey through these pages, you will encounter a wide array of themes and subjects. Some pieces may challenge your preconceptions, while others may inspire you to see the world from a new perspective. Together, they represent the collective wisdom and creativity of our community.

We extend our deepest gratitude to all the contributors who have shared their work with us. Your passion and brilliance are the heartbeat of this collection. We also thank our readers, whose support and engagement are crucial to the continued growth and success of PGR.

It is our hope that "Glimpses of PGR" will not only entertain and enlighten but also spark conversations and connections. May this volume serve as a beacon of the extraordinary potential within us all.

Happy reading!

Contents

1

Impediments in Using Landraces and Wild Relatives in Genetic Enhancement

Abstract

The term "enhancement" was first used by Jones (1983) which according to him can be defined as transferring useful genes from exotic or wild types into agronomically acceptable background.

Keywords: *Landraces, wild relatives*

Introduction

Rick (1984) used the term pre-breeding or developmental breeding to describe the same activity. Thus "genetic enhancement" or "pre-breeding" refers to the transfer or introgression of genes or gene combinations from unadapted sources into breeding materials (FAO, 1996)

However, enhancement has been more popularly adopted by PGR scientists. It is an emerging concept emphasizing the use of plant genetic resources

- To meet the market requirement, plant breeders have to develop improved cultivars.
- Crop improvement has led to narrowing down of genetic base resulting in slower progress (genetic gain) in plant breeding and increased risk of genetic vulnerability

Need for Genetic Enhancement

- In improving the level of **resistance to biotic and abiotic stress.**
- In improving **quality characters** such as fibre length, strength, fineness, maturity and uniformity.
- In developing **early maturing genotypes**, which can fit well in multiple cropping systems.
- In developing plant types suitable to **machine picking** etc.
- Use of germplasm will help in **broadening the genetic base of cultivars** as well as in creating vast genetic variability.

- It will also help in value addition of different genotypes through genetic enhancement

Barriers to Increased Use of CWR in Genetic Enhancement

- Lack of quality data
- Sufficient variation in cultivars
- Germplasm management
- Hybridization difficulty or viability
- Human or financial resources
- Ploidy
- CWR inferiority or drag
- Germplasm exchange
- Existing prebred material underused

Lack Quality Data

- The lack of data on CWR, both phenotypic and genotypic, is the most important barrier to their increased use.
- Approximately 7 million crop accessions are currently stored ex situ in genebanks, of which an estimated 2 million are biologically unique (Commission on Genetic Resources for Food and Agriculture, 2010).
- Much critical information is lacking for a lot of this material.
    - Passport data
    - Characterization
    - Evaluation data

It leads to some unanswered questions

1. How much diversity is in each accession?
2. How similar accessions are?
3. How much duplication exists within and among collections?
4. How to capture the most diversity across the genepool?
5. How to find alleles for specific traits?

Overcome by using standardized systems for identifying accessions and agreed taxonomies are essential measures for improving documentation and will facilitate the development of standardized protocols for managing collections.

Perception of Wild Species' Inferiority Relative to Elite Material

- Wild species can carry beneficial allelic variation for traits without expressing them directly. This "hidden" or "cryptic" variation. sometimes be revealed during the crossing process in transgressive segregation.
- It is difcult to predict potential yield gains from nondomesticated plants with small seeds and pods that shatter
- Identifying interesting alleles when their measured effects are masked by inferior wild background traits.
- Linkage drag

Biological Barrier to Crossing

- Cross incompatibility
- Hybrid inviability
- Hybrid sterility
- Hybrid breakdown

Cross Incompatibility

- This is the inability of the functional pollen grains of one species or genus to effect fertilization in another species or genes.
- There are three main reasons of cross incompatibility viz
    i. Lack of pollen germination,
    ii. Insufficient growth of pollen tube to reach ovule
    iii. Inability of male gamete to unite with the egg cell.

These barriers are known as Pre –fertilization barriers

This is overcome by employing different techniques **like reciprocal crosses, bridge crosses, using pollen mixtures, pistil manipulations, use of growth regulators etc.**

Hybrid Inviability

This refers to the inviability of the hybrid zygote or embryo. In some cases, zygote formation occurs, but further development of the zygote is arrested. In some other cases, after the completion of the initial stages of development, the embryo gets aborted.

The reasons for this are:

- Unfavorable interactions between the chromosomes of the two species.
- Unfavorable interaction of the endosperm with the embryo.
- Disharmony between cytoplasm and nuclear genes.

Overcome by reciprocal crosses, application of growth hormones and embryo rescue are the techniques that can be used to overcome this problem

Hybrid Breakdown

- Hybrid breakdown is a major problem in interspecific crosses.
- When F_1 hybrid plants of interspecific crosses are vigrous and fertile but there F_2 progeny is weak and sterile it is known as hybrid breakdown.
- So hybrid breakdown hinders the progress of interspecific gene transfer.
- This may be due to the structural difference of chromosomes or problems in gene combinations.
- Overcome by growing large population.

Other factors

- Identifying, isolating and transferring traits of interest from wild species into crop background, then evaluating the resulting material, is a time consuming and risky endeavor that require significant funding and long term commitment. It is overcome by Potential strategy to overcome- Public - Private partnership
- Inadequate human resource capacity

Examples

Wheat

In wheat, there are three types of species, viz. diploid (2n = 14), tetraploid (2n = 28) and hexaploid (2n = 42). The cross between common wheat (*Triticum aestivum*, 2n = 42) and durum wheat (*T. durum*, 2n = 28) are partially fertile. In both these species chromosomes of A and B genomes are common and as a result the F_1 hybrids are partially fertile. In F_1 there are 14 bivalents and 7 univalents during meiosis. There is occasional seed set in this cross.

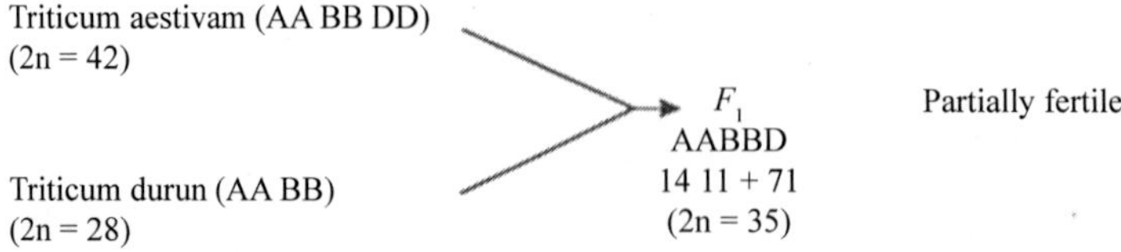

Cotton

In cotton, there are two types of species, viz. diploid (2n = 26) and tetraploid (2n = 52). The cross between American cultivated cotton (*G. hirsutum*, 2n = 52) and American wild diploid (*G. thurberi*) are partially fertile, because these two species have chromosomes of D genome in common. Meiosis in F_1 leads to formation of 13 bivalents and 13 univalents. There is occasional seed set in this cross.

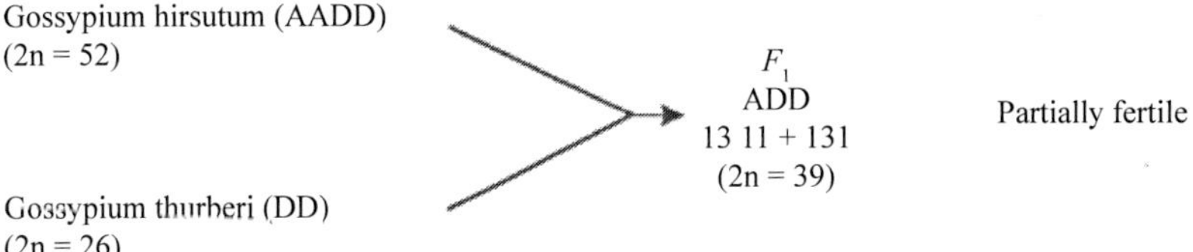

Harland (1940) made a cross between Asian cultivated diploid (*Gossypium arboreum*, 2n = 26) and American wild diploid (*G. thurberi*, 2n = 26). The F_1 was sterile. Treatment of F_1 plants with colchicine resulted in the production of fertile amphidiploid (2n = 52) which was similar to upland cotton (*G. hirsutum*).

Tobacco

In tobacco, there are three types of species, viz. diploid (2n = 24), tetraploid (2n = 48) and hexaploid (2n = 72). The cross between hexaploid wild tobacco (*Nicotiana digluta*) and common tetraploid tobacco (*N. tabacum*) shows partial fertility due to common chromosomes of T_1 and T_2 genomes in these two species. Meiosis in F_1 leads to formation of 24 bivalents and 12 univalents. There is rarely seed set in this cross.

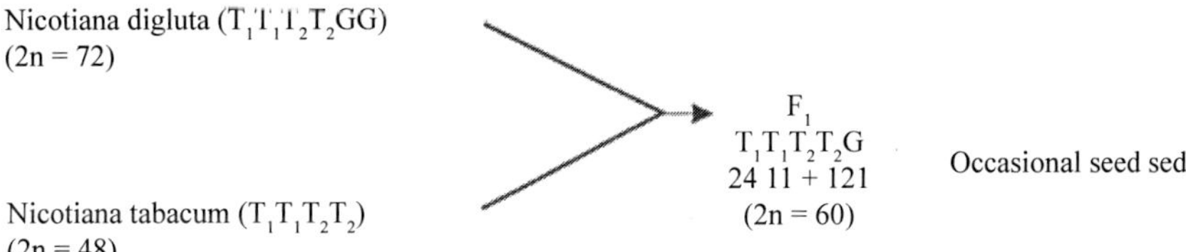

Clausen and Goodspeed (1928) made a cross between two wild diploid species of tobacco, viz. *Nicotiana sylvestris* (2n = 24) and *N. tomentosa* (2n = 24). The F_1 hybrid was sterile. When the F_1 plants were treated with colchicine, a fully fertile tetraploid (2n = 48) was obtained which resembled cultivated species (*N. tabacum*).

They made another cross between another two wild diploid species of tobacco, namely *N. paniculata* (2n = 24) and *N. undulata* (2n = 24). Again the F_1 was sterile. Treatment of F_1 with colchicine resulted in the production of fertile amphidiploid (2n = 48) which was similar to cultivated species *N. rustica.*

Brassica

Interspecific crosses in the genus Brassica were made by several workers. Three crosses were made among three species namely cabbage (*Brassica oleracea*), rapeseed (*B. campestris*) and black mustard (*B. nigra*). The F_1 hybrids were sterile in all the three crosses.

The treatment of F_1 plants with colchicine resulted in the production of fertile amphidiploid in each cross as given below:

Crosses		F_1	New species	Remarks
B. nigra x B. oleracea		Sterile, Colchicine	B. carinata	Fertile
2n=16 (BB)	2n = 18 (CC)	2n = 17 treatment (BC)	2n = 34	BBCC
B. nigra x B. campestris		Sterile, Colchicine treatment	B. juncea	Fertile
2n = 16 (BB)	2n = 20 (AA)	2n = 16 (AB)	2n = 32	AA BB
B. oleracea x B. campestris		Sterile. Colchicine treatment	B. napus	Fertile
2N = 18 (CC)	2N = 20 (AA)	2N = 19 (AC)	2N = 38	AACC

Vigna

Interspecific crosses were made between green-gram (Vigna radiata, 2n = 22) and black-gram (V. mungo, 2n = 22) by Singh and Singh, 1975 and others. The F_1 was sterile. The doubling of chromosome number of F_1 through colchicine treatment resulted in the production of fertile amphidiploid.

References

Bates, L.S., & Deyoe, C.W. (1973). Wide hybridization and cereal improvement. Economic Botany, 27(4): 401-412.

Dempewolf, H., Baute, G., Anderson, J., Kilian, B., Smith, C., & Guarino, L. (2017). Past and future use of wild relatives in crop breeding. Crop Science, 57(3): 1070-1082.

2

Indian Plant Variety Protection and Farmers Right Act & Biodiversity Bill An Analysis

Abstract

As per the TRIPS Agreement(1994) "Members shall provide for the protection of plant varieties either by patents or by an effective sui generis system or by any combination thereof." India has implemented the Protection of Plant Varieties and Farmers' Rights Act, 2001

***Keywords**: PPVFR, TRIPS, Biodiversity bill*

PPVFR act is an effective system for protection of plant varieties, the rights of farmers and plant breeders and to encourage the development of new varieties of plants. Stimulate investment for research and development to produce new plant varieties.Facilitate the growth of the seed industry that will ensure the availability of high quality seeds and planting material to the farmers. Registration of a plant variety gives protection only in India. Provides the exclusive right to produce, sell, market, distribute, import, or export the variety.

Milestones in Seed Sector

- 1957- First AICRP
- 1963 -NSC established
- 1966 -Seed act passed
- 1968- Operational
- 1971- Indian Minimum Seed Certification Standards adopted
- 1975- GSSC established
- 1977- AIC NSP launched
- 1980- GSSCA Established
- 2001- PPVFR Act passed

PPVFR- At a glance

- PPVFR act was included under Act 53 in 2001.
- Rule- 2003
- PPVFR Authority was established (section 3 of the PPVFR Act) during 2005.
- Authority started receiving applications for Registration of Varieties of 12 notified crop species from 21st May, 2007.
- India is the first country to passed legislation granting Farmers' Rights in the form of the PPVFR.
- PPVFR Authority- Administrative apex body
- PVPA Tribunal- Judicial apex body

Administrative Frame Work of PPVFR

- The Chairperson - Chief Executive of the Authority, appointed by GOI

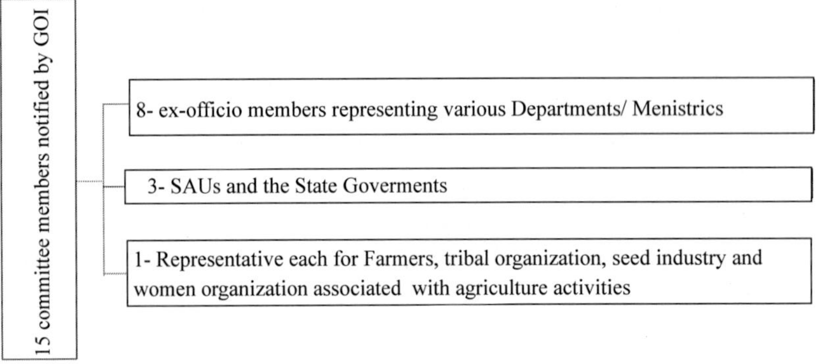

General Functions of the PPVFR Authority

- Registration of new plant varieties, essentially derived varieties (EDV) and extant varieties
- Developing DUS (Distinctiveness, Uniformity and Stability) test guidelines for new plant species
- Developing characterization and documentation of registered varieties
- Compulsory cataloging facilities for all variety of plants
- Documentation, indexing and cataloguing of farmers' varieties
- Recognizing and rewarding farmers, community of farmers.
- Maintenance of the National Register of Plant Varieties and
- Maintenance of National Gene Bank

Overview on PPVFR Act

- 1st Title, definitions
- 2nd PPVFR authority
- 3rd Registration of plant varieties and EDV
- 4th Duration and effect of registration and benefit sharing
- 5th Surrender and revocation of certificate
- 6th Farmer's rights
- 7th Compulsory licence
- 8th PVPA Tribunal
- 9th Finance, accounts, audit
- 10th Infringement, offences and penalties
- 11th Miscellaneous clauses

What can be registered under PPVFR?

- New varieties
- Farmers' varieties
- Extant varieties
- Essentially derived varieties (EDV)

What can not be registered under PPVFR?

- Prevention of commercial exploitation of those variety which is necessary to protect public morality to human, animal and plant health and cause harm full effect to environment.
- Species involving any technology viz., Genetic use restriction technology (GURT) or Terminator Gene Technology.

Eligibility/Criteria for registration

- **Novel:** Should not have been offered for sale before one year from date of filing of application (India), 6 years in case of trees and vines(Outside India), other case-4 year.
- **Distinct:** Should be clearly distinguishable from any other variety by at least one essential characteristic.
- **Uniform**: Should be sufficiently uniform in its essential characteristics.
- **Stable**: Essential characteristics should remain unchanged after repeated propagation.

Status of PPV at different countries

Country	Legislation	Scope of coverage	Plant variety patents
China	Regulations of the People's Republic of China on the Protection of New Varieties of Plants (1999).	41 crops currently eligible. Certificates have been issued for 15 species through 2004.	Hybrids may fall under scope of patents for a breeding
India	Protection of Plant Varieties and Farmers' Rights Act (2001) establishes PVP. Implementation beginning in 2005.	No crops excluded, but exemption for varieties whose commercial exploitation would be a danger to public order, public health, etc.	No patents of plant varieties allowed
Uganda	Draft Plant Variety Protection Act is still before Parliament.	No crops excluded in draft bill	No patents of plant varieties allowed

DUS test under PPVFR

- Application along with sufficient quantity of seed should be submitted to the Registrar.
- Evaluate Seeds of a variety along with parental material to conform the standards as may be prescribed.
- Keep safe to unaltered viability and quality of seeds.
- Applicant shall deposit the fees.(DUS+special test) (except farmer)
- A part of the seed sample - to conduct of DUS tests.
- Remaining part – To Maintain seed samples at National gene bank for their entire period of protection by the Authority.

Duration & effect of registration and benefit sharing of protected variety

Crop	Initial	Extendable	Total
Field crop	6	9	15
Trees and Vines	9	9	18
Extant variety	-	-	15 years from date of notification
Any other case	-	-	15

National Gene Fund

Sources of funds

- Compensation amount deposited
- Annual fees as royalty
- Benefit sharing amount deposited
- Contribution from National or International organization

Uses of funds

- Reimbursement of benefit sharing
- Reimbursement of compensation
- Supporting Conservation of genetic resources
- Any other activity indicated in Act

Plant Variety Protection Appellate Tribunal under PPVFR

- Central Govt. constitutes Plant Variety Protection Appellate Tribunal.
- It consists of Chairman, Judicial members and technical members (Plant Breeding and Genetics).
- Works on jurisdiction, powers, and authority conferred under this act.
- The decision of the tribunal under this act is executable as a judgment of Civil Court.

Provision of rights under PPVFR

A. Breeders' rights

- Rights to produce, sell, market, distribute, import or export the protected variety.
- Can appoint agent/licensee and exercise civil remedy in case of infringement of right.

B. Researchers' rights

- Permitted one time use of registered variety.
- Can use any of the registered variety (initial source) for conducting experiments to develop another variety.
- But repeated use of such variety as parental line needs prior permission of the registered breeder.

C. Farmers Rights

1. Farmers' Right on Seed: saving ,sowing, exchange, sharing, selling to other farmers (not branded seed)
2. Farmers Rights' to register Traditional varieties
3. Farmers' Right for rewards and recognition
4. Farmers' right for benefit sharing
5. Farmers' Right to get compensation for the loss suffered from registered variety
6. Farmers' Right to receive the compensation of undisclosed use of traditional varieties

7. Farmers' Right to receive the compensation of undisclosed use of traditional varieties
8. Farmers' Right for the seeds of registered variety
9. right for receiving the free service under the Act
10. Farmers' Right for protection against innocent infringement

Comparison of PPVFR Act, 2001 and Seed Bill, 2004

S.No.	PPV & FR Act, 2001	Seed Bill, 2004
1	Farmer get compensation from the PPV authority which is all the more simpler	Farmer has to claim compensation from a consumer court and redresses under the Consumer Protection Act, 1986
2	Requires the declaration of origin of variety along with pedigree details	Does not require the declaration of origin of variety along with pedigree details
3	provides rewards for farmers contribution and also the benefit sharing	Does not grant any recognition to the contribution of farmers
4	Provides compulsory licensing which safeguards the interests of farming community to ensure adequate seed supply at reasonable price on the Government.	Seed dealers are not under any obligation to provide reasonable seed supply to farmers

Biodiversity Bill, 2000

- A Bill to provide for conservation of biological diversity, sustainable use of its components and equitable sharing of benefits arising out of the use of biological resources in India.
- It was enacted by Parliament in the Fiftieth Year of the Republic of India.

Necessity of Biodiversity Bill, 2000

- Secure national legal commitment for protection of biological diversity and the maintenance of ecological processes and systems
- Establish a national legislative framework for the conservation of biological diversity and the ecologically sustainable development of its components
- Ensure that safe biodiversity conservation standards for decision-making and activities are established as a means of operationalising the precautionary principle in order to anticipate, prevent and attack the causes of significant reduction and loss of biological diversity in the face of scientific uncertainty
- Prevent species and ecological communities from becoming vulnerable and to protect critical habitat
- Codify national institutional and administrative arrangements necessary to implement the United Nations Convention on Biological Diversity

- Establish awareness to care for the environment among all natural resource owners, managers, users and others whose actions could foreseeably harm the environment;
- Ensure adequate funding for achievement of the above objects

Applicability

- All varieties of life-forms including plants and animals and micro-organisms belonging to all general species, wild or cultivated, occurring naturally or modified in any manner through any process, in relation to their cell lives, generic material, characteristics, traits, products and the processes involved therein
- Any level of continental, pelagic, coastal or insular biological organization that is found in national territory and in the waters under its jurisdiction

Various important chapters of Biodiversity bill in brief

Chapter II: Establishment And Constitution Of National Authority

- NBA having perpetual succession and a common seal with power to acquire, hold and dispose of property, both movable and immovable, and to enter into contract
- The Chairperson shall be the Chief Executive Officer of the National Authority
- Head office of NBA should be at Delhi.

Chapter III: Responsibilities of the Central Government for conservation and sustainable use

- The Central Government shall be in consultation with the National Authority.
- Develop national strategies, plans or programmes for the consevation and sustainable use of biological diversity and to adapt for this purpose existing strategies, plans or programmes.
- Promote programmes oriented towards the improvement of the methods of production, conservation and distribution of foods with the full utilization of national and international technical and scientific knowledge which is environmentally and culturally clean and secure and shall also work towards perfecting the present agrarian regime consistent with the objectives of social and economic justice.
- Prohibit grant of patents or other individual intellectual property rights on cultivated species or substances improved on breeding farms which are used as an alimentary or medicinal base.

Guidelines for benefit sharing biological resources

- NBA forms guidelines of benefit sharing of biological resources consistent with the objectives of this Act, without adversely affecting the rights of the local people.
- Identify and declare bio-regions and protected areas and to adopt the precautionary principle.
- Regulate access to biological diversity on a national basis, providing inter-alia for prior informed consent of the "owners" of biological resources.
- Notification of areas of biodiversity importance as Biological Diversity Heritage Sites
- Notification of threatened species and prohibit or regulate their collection for any purpose and take appropriate steps to rehabilitate and preserve such species.
- Development of repositories for different categories of biological resources

Chapter VII: National State And Local Biodiversity Fund

Used for

- Channelling benefits to the conservers of biological resources, or creators and holders of knowledge.
- Conservation of biological resources and in particular, conservation and development of biological resources in areas from where such resource, or knowledge has been accessed;
- Socio-economic development of such areas in consultation with the concerned Panchayat or Municipality;
- Conservation of Heritage Sites notified under section 20.

Chapter VIII: Finance, accounts and audit

Chapter IX: Supersession of dissolution of authority

Chapter X: Miscellaneous

References

Anandhi, K. (2015). Protection of plant varieties and farmer's rights (PPVFR) act: A review. Current Advances in Agricultural Sciences (An International Journal), 7(2): 101-105.

Tonapi, V. A. (2008). PPV and FR Act and National Seed Policy. Sorghum Improvement in the New Millennium. Patancheru 502 324, Andhra Pradesh, India: International Crops Research Institute for the Semi-Arid Tropics. 340 pp, 48, 232.

3

PGR and Genomics for Crop Improvement

Abstract

Wide hybridization: *Cross between a crop plant and a related wild species (a wide cross) enables* **transfer of valuable genes** *from the wild species into the background of crop plant.* **Exploitation of heterosis** - *If two distinct inbred plants were crossed together, the resulting progeny (called F1 hybrids) exceeded both parents in vigor and yield. This phenomenon, known as* **hybrid vigor or heterosis**, *is the basis of a large hybrid seed industry for maize, sorghum, sunflower, and many vegetables.*

Keywords: *Crop, wildspecies, vigor and yield.*

Mutation: Natural genetic variation has exhausted. **Genetic variation** can be induced by *mutations* or changes in the DNA sequences of the plants.

Additional Tools for Plant Improvement

Increase in yield with no additional lands available for cultivation depends on development of high yielding varieties suitable for marginal cultivation.

- Food security
- Nutritional security

Need for additional tools to increase efficiency of breeding methods for genetic enhancement and propagation.

- Genomics

Genomics – Definition

Genomics - Study of an organism's **entire genome / full genetic complement** of an organism.

Genomics classified into.

1. **Structural genomics** - genetic structure of each chromosome of the genome.
   - Mapping
   - Sequencing

2. **Functional genomics** - function of all genes present in the entire genome (differential expression in a particular cell/tissue at a particular time)
   - Transcriptomics - complete set of RNAs transcribed from a genome
   - Proteomics – complete set of proteins encoded by a genome
3. **Comparative genomics** - genomic features of different organisms are compared.

PGR and genomics

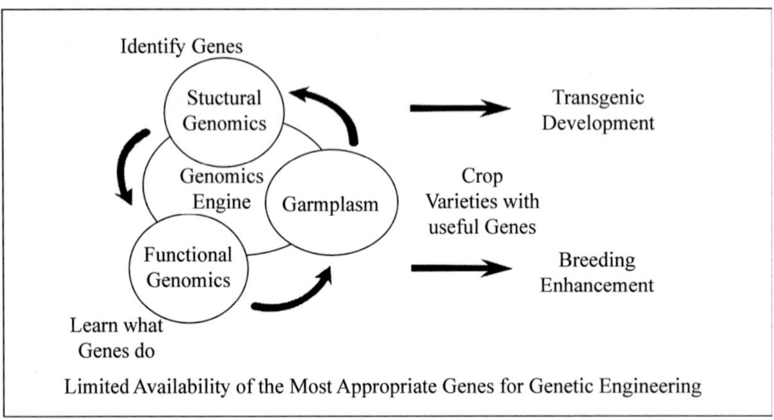

Limited Availability of the Most Appropriate Genes for Genetic Engineering

Genomic technologies in germplasm utilization

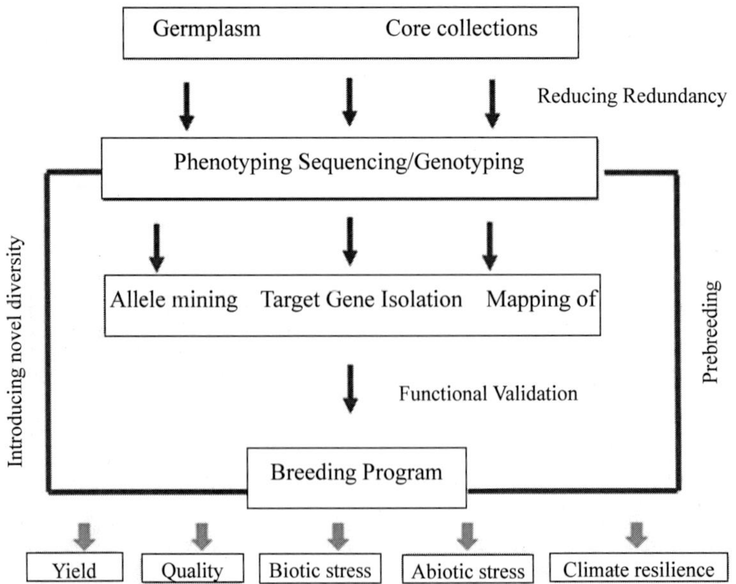

1. Mapping of genes/ QTLs

A large number of monogenic and polygenic genomic loci for various traits have been identified. DNA markers are used to mark a genomic region, tag a gene or distinguish individuals from each other. Molecular markers have facilitated construction of physical and genetic maps of genes/QTL on the genome.

Applications

I. **Marker Assisted Selection** - Scan new varieties to directly select for the presence of the markers and indirectly select for the desired genes. MAS subverts the need for phenotypic selection which can be difficult, time consuming , costly and influenced by the environment. MAS can reduce the breeding cycle by about 4 years which would amount to saving $ 200 million in 25 years compared to conventional breeding technique.

MAS in different crops

Crop	Trait
Rice	Blast resistance, bacterial blight resistance, brown planthopper resistance, submergence, salt, drought, cold tolerance, cooking quality, eating
	quality, yield, heading date, genetic male sterility, basal root thickness, root traits
Wheat	Fusarium head blight (FHB) resistance, leaf rust resistance, powdery mildew resistance, stripe rust resistance, cereal cyst nematode resistance, glutenin
	quality, preharvest sprouting tolerance (PHST), grain protein content, dough properties, durable rust resistance and height
Barley	Barley yellow mosaic virus I-HI resistance, cereal cyst nematode resistance, barley stripe rust resistance, leaf rust resistance, loose and covered smut resistance, malting quality, yield
Soybean	Soybean mosaic virus (SMV) resistance, resistance to frogeye leaf spot (Cercospora sojina), ear-worm resistance
Whitebean	Bean golden yellow mosaic virus (BGYMV), common bacterial blight resistance
Tomato	Black mold resistance, acyl sugar mediated pest resistance, bacterial spot and speck resistance, fruit quality
Pepper	Tobaino virus resistance, tomato spotted wilt virus resistance, root knot nematode resistance, po-tyvirus reistance
Cucumber	Yield contributing traits, multiple lateral branching
Maize	Cora borer resistance, seedling emergence, quality protein maize (QPM), earliness and grain vield
Pearlrmllet	Disease resistance and grain yield
Potato	Root-knot nematode resistance, potato virus X and Y resistance,root cyst nematode. wart resistance

C'otton	Fiber strength
Broccoli	Resistance to diamond lack moth
Dry bean	Sclerotinia white mold resistance

II. **Introgression/Backcrossing (MABB)** - Gene introgression for a desirable trait (like resistance to a specific disease) from donor plant (Wild type) into elite cultivar by repeated backcrossing till maximum recovery of recurrent parent genome. Conventional backcross breeding takes about 6-8 generations with problem of linkage drag, which can be overcome by MABB.

III. **Gene Pyramiding-** Pyramiding is the accumulation of several desired alleles into a single line or cultivar (background) using markers.

Case study: Development of Co43 (Sub1) rice by MABB (Rahman *et al.*, 2018)

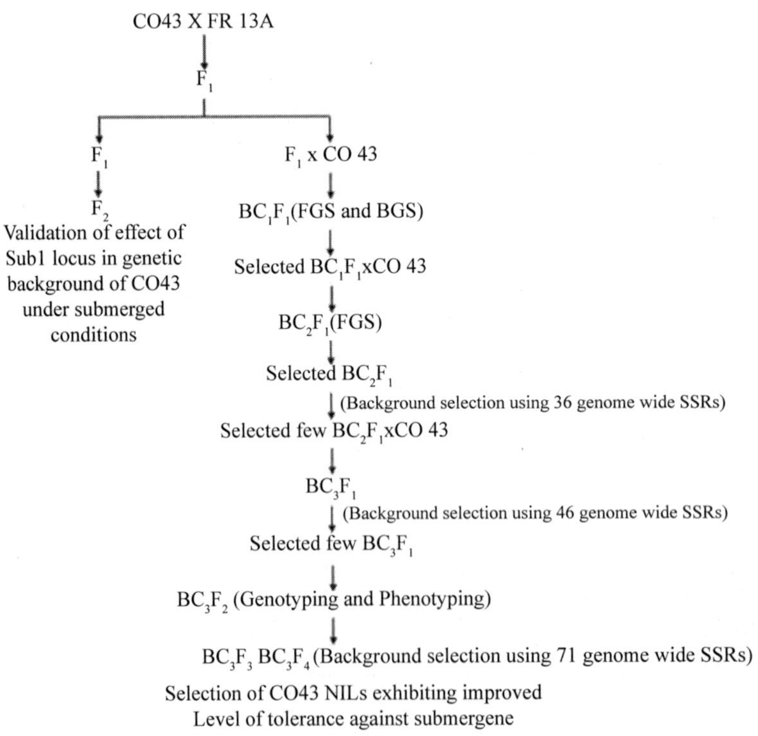

Advances

Modern Breeding Approaches	
Association Mapping	Utilizes natural variation and linkage disequilibrium existing in a natural population or genii-plasm to identify
	genomic regions associated with a trait. Offers advantages of increased QTL resolution, reduced research time
	and an increased sampling of molecular variation, over traditional linkage mapping. The advent ofNGS
	technologies has enabled both high throughput candidate gene and genome wide association mapping.
Advanced Backcross-QTL analysis	Involves simultaneous discovery-' mapping and transfer of QTLs from a wild species to a cultivated variety for
	the improvement of the trait. BCiFj or BC'2F3 population is analysed for the presence of the target trait and
	is also genotyped using molecular markers,
Genome wide selection (GWS or GS)	Used to predict the breeding values of lines in a population by analyzing their phenotypes and high-density marker scores, the latter is used to calculate genomic estimated breeding values (GEBV) as the sum of the effects of all QTLs across the genome. These values are then used to select the progeny lines for advancement in the breeding cycle

2. Genome sequencing and re-sequencing

The inexpensive sequencing and resequencing technologies are the major driving forces behind increased number of assembled plant genomes of different crops including wild relatives.

Next Generation Sequencing

- **Short read** (Illumina) – read lengths shorter than 400 bp – helps to capture SNP and small InDel variation
- **Long read** (PacBio) – read lengths of several thousand bp of continuous sequence – helps to sequence complex genomic regions (repetitive elements) – helps to identify large insertions, deletions and rearrangements
- Whole genome shotgun re-sequencing to a high genome coverage/depth
- Whole genome shotgun sequencing to a low genome coverage
- RNAseq – Whole genome expression profiling

Whole-genome resequencing

- Single reference genome doesnot represent the total diversity within a species. Resequencing of cultivars, landraces and wild accessions is required to harness the total genetic variation and to **identify novel**

variations (superior alleles) for a large number of genes through genotype-phenotype associations

- Discovery, validation, and assessment of diagnostic markers in different crops and it provides **genome-wide markers**
- Large collections of **functional markers** which enhance gene assisted breeding, reducing the possibility of losing the desirable trait variation due to recombination.
- Determines process of origin, domestication, population structure, hierarchical clustering and identifies lines with deleterious mutations
- **Genotyping by sequencing (GBS)** approach in combination with phenotyping can be used for **identification of QTLs** controlling various traits

Pangenomics

To capture the entire genomic sequence present within the species, including the complete gene set, the pangenome needs to be sequenced.

Pangenome is the sum of all the genes of a particular species rather than single sample reference. It is a better representation of diversity by reducing sampling bias. It helps to identify/study genomic structural diversity (structural variations) - differentiating genomes via. presence and absence of sequences known as presence/absence variants (PAV) and differences in copy numbers known as copy number variants (CNV).

Known biological processes influenced by CNVs/PAVs are metabolite production, flowering time, submergence tolerance, phosphorus uptake and biotic stress response. Pangenome sequence assembly and linking of these structural variations to the key traits in crops has accelerated the prebreeding programs.

Example: Opium poppy contains a 10 gene cluster which displays PAV, being only present in plants producing noscapine an antitumor alkaloid. Investigation of wheat genes that regulate flowering by altering photoperiod response (Ppd-B1 alleles) or vernalization requirement (Vrn-A1 alleles) revealed that both of those genes display CNV. Alleles with an increased copy number of Ppd-B1 confer an early flowering day neutral phenotype, while plants with an increased copy number of Vrn-A1 have an increased requirement for vernalization so that longer periods of cold are required to potentiate flowering.

3. Allele Mining of Candidate Genes

Dissect naturally occurring allelic variants / new haplotypes of candidate genes with superior agronomic qualities. It is searching for useful alleles of genes from a wide range of cultivars, related species and even across species.

Two approaches for allele mining: (i) eco-TILLING (targeting induced local lesions in genomes), (ii) re sequencing or sequencing based allele mining

Applications - identification of key alleles conferring:

- Resistance to biotic stresses
- Tolerance to abiotic stresses
- Greater nutrient use efficiency
- Enhanced yield
- Improved quality

Eco – TILLING - TILLING (Targeting Induced Local Lesions IN Genomes) is a technique that can identify single base-pair allelic variation in target gene (more specifically induced point mutations) while Eco-Tilling technique detects natural mutation. EcoTilling also relies on the enzymatic cleavage of heteroduplexed DNA (formed due to single nucleotide mismatch in sequence between reference and test genotype) with a single strand specific nuclease (i.e., Cel-1, mung bean nuclease, S1 nuclease, etc.) under specific conditions followed by detection through Li-Cor genotypers. At point mutations, there will be a cleavage by the nuclease to produce two cleaved products whose sizes will be equal to the size of full length product. The presence, type and location of point mutation or SNP will be confirmed by sequencing the amplicon from the test genotype that carry the mutation.

Sequencing based allele mining - involves amplification of alleles in diverse genotypes through PCR followed by identification of nucleotide variation by DNA sequencing techniques. Sequencing-based allele mining would help to analyze individuals for haplotype structure and diversity to infer genetic association studies in plants. Unlike EcoTilling, sequencing-based allele mining does not require much sophisticated equipment or involve tedious steps, but involves huge costs of sequencing.

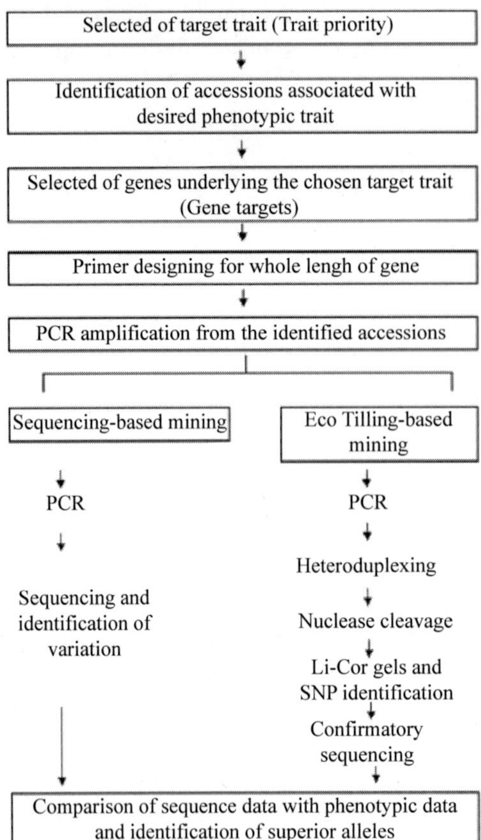

Case study: Haplotype analysis of key genes governing grain yield and quality traits (Abbai *et al.*, 2019)

Haplotype analysis of 120 previously functionally characterized genes, influencing grain yield (87 genes) and grain quality (33 genes) revealed significant variations in the 3K rice genome (RG) panel. It revealed that more than 75% of the genes had haplotypes ranging from 2 to 15 among the 3024 lines.

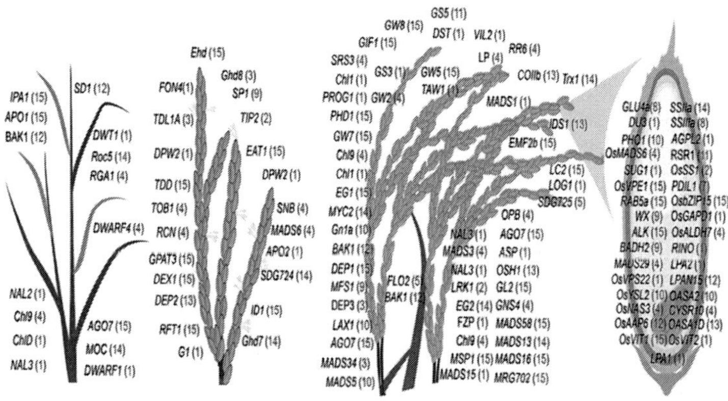

The tailored rice with superior haplotypes for grain yield and quality. The findings of this study could be employed to develop a designer rice genotype comprising superior haplotype combinations of the target genes such as MOC1-H9 & IPA1-H14 for higher tiller number, Ghd7-H8 & TOB1-H10 for early, Ghd7-H14 & OsVIL3-H14 for medium duration, Ghd7-H6, SNB-H9 & TRX1-H9 for late flowering, DEP3-H2, DEP1-H2 & SP1-H3 for long panicles, SD1-H8 for semi-dwarf nature, LAX1-H5, OSH1-H4 & LP-H13 resulting in increased panicle branching, PHD1-H14, AGO7-H15 & ROC5-H2 for high yield, along with GS5-H4 for slender, GS5-H5 & GW2-H2 for medium slender, GS5-H9 for bold grains, RSR1-H8 for intermediate amylose content and OsNAS3- H2 for increased Fe and Zn concentration in grains. B, Bold grain; MS, Medium slender type grain; S, Slender type grain.

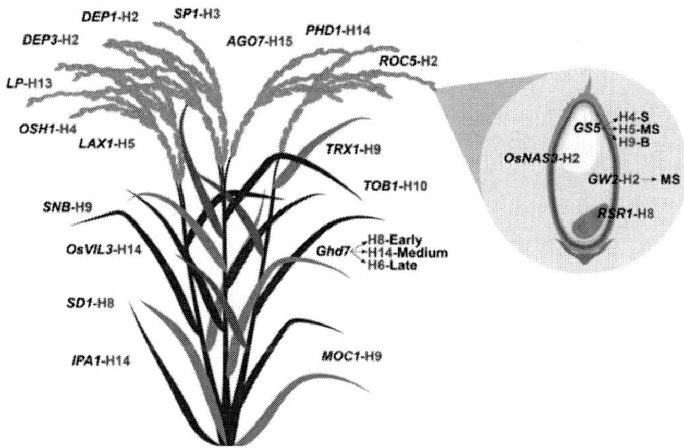

4. Genetic engineering (Transgenesis)

Plant genetic engineering comes into play when natural genetic variation/ crop diversity for a particular trait has exhausted. In such cases we adopt genetic engineering for the transfer and random insertion into the host plant genomes of genes isolated from other plant species or from other organisms which is not otherwise possible by natural pollination or other conventional hybridization approaches. Mostly these are applied for gene transfer between genera or between different kingdoms. Knowledge of the structure and function of plant genomes in major agricultural and related PGR aims to the isolation of genes underlying the characters of interest and their precise modification or transfer into targeted varieties.

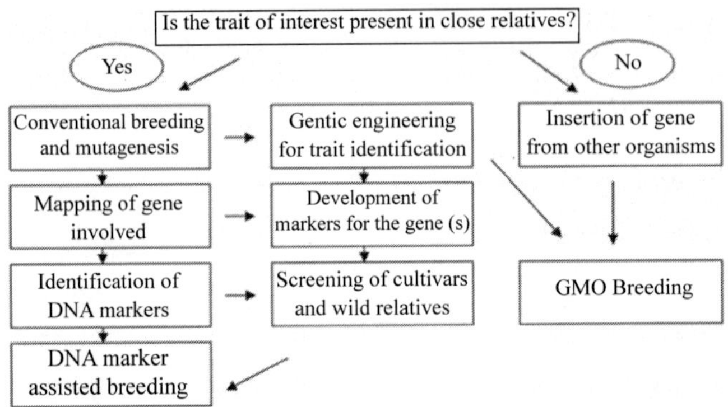

5. Cisgenesis/Intragenesis

Cisgenesis - A cisgene is a natural gene, coding for a trait, from the crop plant itself or from a sexually compatible donor plant. Inserted gene is unchanged with its own introns and regulatory sequences in the normal sense orientation.

Intransgenesis - in-vitro rearrangement of gene elements. Inserted DNA can be a new combination of DNA fragments from the species itself or from a cross-compatible species in sense or antisense orientation.

Advantages: Lack of linkage drag and no recurring back crossings required; higher knowledge of transferred sequences.

Crop	**Method**	**Type**	**Gene**	**Trait**
Potato	intragenesis	Silencing	GBSS	High amylopectin
Potato	cisgenesis	Gene from related species	R-genes	Late blight resistance
Apple	cisgenesis	Gene from a related species	*HcrVf2*	Scab resistance
Apple	intragenesis	Gene from a related species	*HcrVf2*	Scab resistance
Grapvine	cisgenesis	Gene from a related species	*VVTL-1*	Fungal disease resistance
Poplar	cisgenesis	Over-expression	Growth related	Plant architechture
Alfalfa	intragenesis	silencing	Cornt	Reduced levels of lignin

6. Genome Editing

GE includes a set of techniques that allow to edit, delete, replace or insert, in a targeted site, specific genomic sequences of interest. They are based on the induction of cuts in double-strand DNA (DSB, double-strand breaks), which are then 'repaired' with two different processes: the non-homologous end-joining (NHEJ) or homology-directed repair (HDR).

The breaks in double-strand DNA can be induced by four systems based on specific enzymes:

- Meganucleases,
- Zinc finger nucleases (ZFN),
- Transcription activator-like effector nucleases (TALEN) and
- Clustered regular interspaced short palindromic repeats/CRISPR-associated nucleases (CRISPR/Cas).

Targeted genome modifications include

i. the induction of mutations at preselected loci to disrupt the function of one or more specific genes;

ii. the editing of existing sequences to reproduce ancient alleles or to introduce novel alleles;

iii. the introduction of new genetic material into specific loci or regions of the genome.

iv. It is also possible to change DNA modifications, such as methylation, in order to modulate gene expression.

Table 2: Examples of agronomic traits modified through the application of genome editing approaches in various crops

Crop	Trait	Cjene	Technology	References
Gene knockout (SDN-I)[1]				
Rice	Resistance to bacterial blight	OsSWEET14; OsSWEETll	TALEN; CRISPR/ Cas9	Li et al. (2012); Zhouet al. (20l5a)
	Fragrance	OsBADH2	TALEN	Shan et al. (2015)
Bread wheat Maize	Resistance to powdery mildew Phytate biosynthesis Leaf epicuticular wax composition Leaf development; Male fertility; Herbicide resistance	TaMLO-Al, TaMLO-Bl, TaMLO-Dl IPK1 glassy? (g!2) UG1; Ms26, Ms45; ALSl, ALS2	TALEN ZFN TALEN CRISPR/ Cas9	Wang et al. (2()14) Shufcla et al. (2009) Charet al. (2015) Svitashev et al. (2015)

Soybean Poplar Potato	Profile and unsaturation level of seed fatty acids Lignin content; Condensed tannin content Accumulation of reducing sugars after cold storage and acrylamide after high-temperature processing Accumulation of steroidal glycoalkaloids	FAD2-1A and FAD2-1B 4CL1; 4CL2 VInv S1SSR2	TALEN CRISPR/Cas9 TALEN TALEN	Haunet al. (2014) Zhou et al. (2015b) Clasen et al. (2015) Sawai et al. (2014)
Tomato	Plant development Leaf development	PROCERA (PRO) ARGONAUTE7 (SIAGO7)	TALEN CRISPR/Cas9	Loret al. (2014) Brooks et al. (2014)
Gene editing (SDN-2)				
Maize Soybean Tobacco	Herbicide resistance Herbicide resistance Herbicide resistance[2]	ALS2 ALSl ALS SuRA and SuRB	CRISPR/Cas9 CRISPR/Cas9 ZFN	Svitashev et al. (2015) Li et al. (2015) Townsend et al. (2009;
Gene replacement/stacking (SDN-3)				
Maize	Phytate production/ herbicide resistance Herbicide resistance”	1PK1/PAT PAT	ZFN CRISPR/Cas9	Shufcla et al. (2009)

GE approaches can be used to modify genes with defined quantitative trait nucleotides (QTNs) that cause a sizeable phenotypic effect. Further, GE tools can be used for broadening the allele pool through generating targeted variations useful for genomic selection.

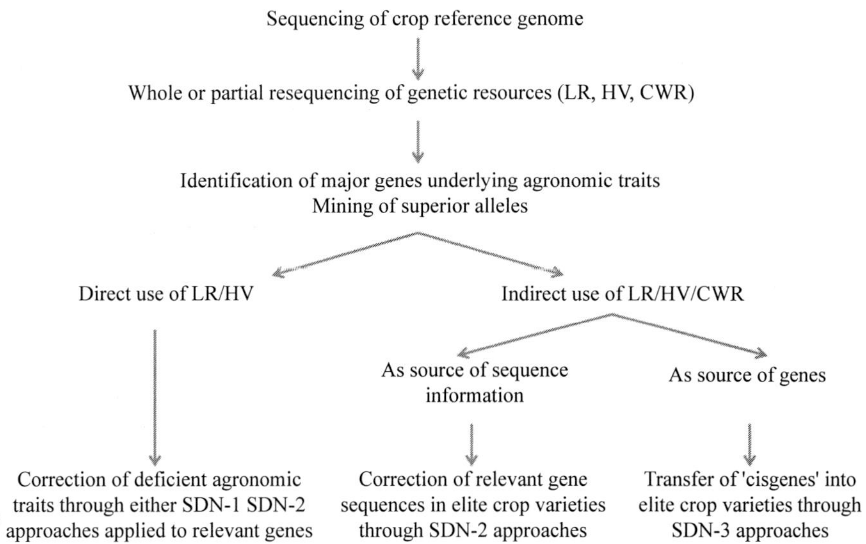

LR: landraces; HV: heirloom varieties; CWR: crop wild relatives; SDN1 – InDels, SDN2 – Gene replacement, SDN3 – Gene Insertion

References

Abbai, R., Singh, V. K., Nachimuthu, V. V., Sinha, P., Selvaraj, R., Vipparla, A. K., & Kumar, A. (2019). Haplotype analysis of key genes governing grain yield and quality traits across 3K RG panel reveals scope for the development of tailor-made rice with enhanced genetic gains. Plant Biotechnology Journal.

Cardi, T. (2016). Cisgenesis and genome editing: Combining concepts and efforts for a smarter use of genetic resources in crop breeding. Plant Breeding, 135(2): 139-147.

Rahman, H., Dakshinamurthi, V., Ramasamy, S., Manickam, S., Kaliyaperumal, A. K., Raha, S., and Raveendran, M. (2018). Introgression of submergence tolerance into CO 43, a popular rice variety of India, through marker-assisted backcross breeding. Czech Journal of Genetics and Plant Breeding, 54(3): 101-108.

Singh, K., Kumar, S., Kumar, S. R., Singh, M. and Gupta, K. (2019). Plant genetic resources management and pre-breeding in genomics era. Indian J. Genet, 79(1 Suppl 117): 130.

Yadav, S., Kumar, S., Savita, K., Singh, A.K. and Singh, R. (2015). Genomics for Crop Improvement–an Indian perspective. Biotech Today: An International Journal of Biological Sciences, 5(1): 45-50.

4

Genetic Enhancement / Pre-breeding of Crop Improvement Including Hybrid Development

Abstract

The narrow genetic base of cultivars coupled with low utilization of genetic resources is the major factor limiting production and productivity globally. To exploit this genetic diversity pre-breeding offers a unique opportunity by introgression of desirable genes from wild germplasm into cultivated backgrounds readily used with minimum linkage drag. Pre-Breeding term was first coined by Rick in 1984. Alternative term is "Genetic enhancement" coined by Jones in 1983.

Keywords: *Genetic enhancement,Germplasm, Pre breeding*

Introduction

It refers to all activities designed to identify materials that cannot be used directly in breeding programmes, and further to transfer these traits to an intermediate set of materials that breeders can use further in producing new varieties for farmers. It is a bridge between Genetic Resources and Crop Improvement. That is collaboration between germplasm curator and plant breeder.

Crosses between adapted and exotic materials, where different proportions of introgression are obtained and evaluated, have been denominated as semi-exotic materials. That should be readily used in regular breeding programme for cultivar development.

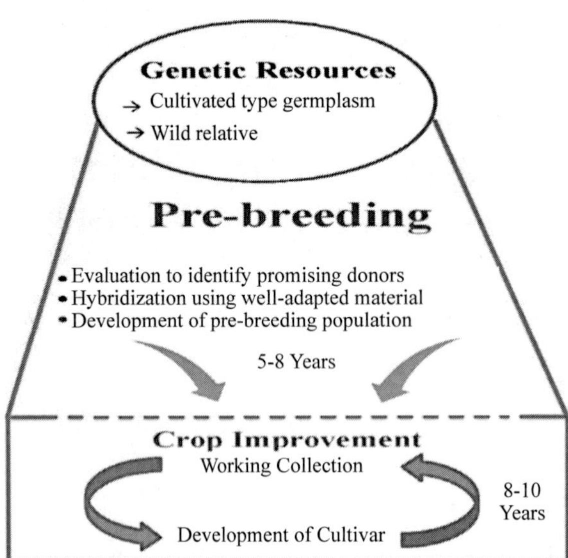

Objectives of pre-breeding

- Improved germplasm and associated genetic knowledge that enhance resistance expression and diversity.
- Reduce genetic uniformity in crops through the use of a wider pool of genetic material to increase yield, resistance to pests and diseases, and other quality traits.
- Identification of desirable traits/genes and their subsequent transfer into a suitable set of parents for further selection.
- Improved parental stocks which can be readily utilized within breeding programs and improved selection methodologies.
- Identify potentially useful genes in a well-organized and documented gene bank Designing strategies that lead to development of an improved germplasm that are ready to use in varietal development.

Why Pre-breeding is required

- Current limited genetic base of agriculture today is apparent a threat to food security.
- Reduction of Biodiversity: genetically uniform modern varieties are replacing the highly diverse local cultivars and landraces in traditional agro-ecosystems.
- Genetic uniformity: Increases genetic vulnerability for pests and diseases.

- The effects of climate change: search for new genes/traits for better adaptation.
- Evolving pest and pathogen populations: motivating plant breeders to look for new sources of resistance in gene banks.

The decision to do pre-breeding is based on the expected efficiency and efficacy of ultimately moving the target traits into cultivars for farmers and source of desired gene(s). Pre-breeding is necessary, if desired genes are available only in one of the following:

1. Gene bank accessions those are not well-adapted to the target environment.
2. Closely related wild species that are easily crossed with the crop species.
3. More distant wild species which are more difficult to cross.

Requirement for pre-breeding

- Collection of underrepresented diversity with detailed passport data.
- Trait mining based on eco-geographic data.
- Coordinated evaluation across environmental conditions to better understand GxE interactions.
- Sharing of improved information from phenotypic and genotypic data between the conservationists, pre-breeders, breeders, and end users.
- Enhanced coordination between basic and applied research communities.

Various activities involved in Pre-breeding

- Characterization of landrace populations.
- *Creation of new parent populations from landrace populations:* Parents with high specific or general combining abilities are selected via progeny testing through well-designed recombination. Progeny testing is performed in a set of target and representative environments with half-sibs, full-sibs, test crosses or recombinant inbred.
- Introgression of new traits from other useful sources.
- Creation of novel traits through mutation
- *Creation of polyploidy*: Developed through hybridization, plant materials exposed to chemical (Colchicine) and physical (high temperature and X-ray) treatments.
- *Acquisition of new information on crop genetics*: Understanding the candidate genes and the pattern of inheritance of the genes in controlling these characters is profoundly significant for effective transfer and to improve the efficiency of selection in cultivar development

- *Development of new plant breeding techniques*: Molecular marker technologies, effective gametocides and cytoplasmic sterility systems with a desired genetic background, double haploids etc.

Major approaches for use of PGR in genetic enhancement

a) **Introgression:** Introgression is transfer of one or more genes from exotic/un-adapted / wild stock to adapted breeding populations. This is achieved by making crosses between the donor and the recurrent parent. The concept of introgression through backcross was evolved by Dr. Edgar Anderson and in cotton it was first visualized by Knight (1945). This method leads to accumulation of genes resulting in enhanced level of genetic expression for the trait. Three methods are followed:

   1. **Recurrent backcross:** This method involves successive backcrossing of the cross made between donor and the recurrent parent with the recurrent parent with or without selection. With this method, the monogenic or oligogenic traits with high heritability can be easily transferred.

   2. **Inbred backcross:** This method involves a limited number of backcrosses (usually one or three). This is followed by several generations of selfing. This method was first given by Wehrhahn and Allard, 1965. This method results in the development of a population of more than 50% of lines that are homozygous and have a common genetic background similar to the recurrent parent. This method has been used to transfer quantitative traits. e.g. improvement of nitrogen fixation and seed protein content in common bean (Bliss, 1985).

   3. **Congruity backcross:** This method was originally proposed by Haghighi and Ascher, in 1988. In this method, backcrossing is done to both donor and recurrent parent in alternate generations. This method was used to recover progeny with increased fertility from *Phaseolus vulgaris* and P. *acutifolius* crosses. It was possible to obtain progenies with intermediate morphologies. This led to effective recombination through the increased levels of heterozygosity.

b) **Incorporation:** Incorporation refers to a large scale programme aiming to develop locally adapted population using exotic / un-adapted germplasm. This was first suggested by Simmonds (1993). In contrast to introgression, incorporation aims at indexing the crop genetic base. The following are the genetic principles of incorporation.

   a. Use of material covering wide range of variability.

   b. Use of un-adapted introduced material.

c. The process is complementary to conventional breeding.
d. The breeding methods will depend on the biology of the crop, its breeding system and reproduction behavior.
e. Maximizing recombination through cyclic or recurrent crossing.
f. Testing for adaptability under diverse agroclimatic conditions.
g. Local genetic adaptation - horizontal resistance (HR) to disease.
h. The outcome of an effective base-broadening programme will be enhanced genetic variance in economic characters and either good material *per se* or good parents for crossing into established programmes.

c) **Wide cross:** A cross of two individuals belonging to different species or different genera is known as wide cross. Such a cross can be (rarely) realized in nature – origin of new species and synthesis of new base populations.

d) **Decentralized participatory plant breeding:** A decentralized-participatory plant breeding program also function with the same line provided with some differences like most of the process takes place in farmers' fields, the decisions are taken jointly by the farmers and the breeder and the process can be implemented at a number of locations involving a large number of farmers evaluating different breeding materials.

e) **Marker Assisted Breeding:** Breeding methods based on DNA molecular marker patterns instead of, or in addition to, their trait values. It is a tool that can help plant breeders select more efficiently for desirable crop traits. When molecular markers are available, conveniently co-segregating with candidate genes, marker-assisted selection (MAS) or marker-aided selection may improve the efficiency of selections of simple traits in conventional plant breeding programs.

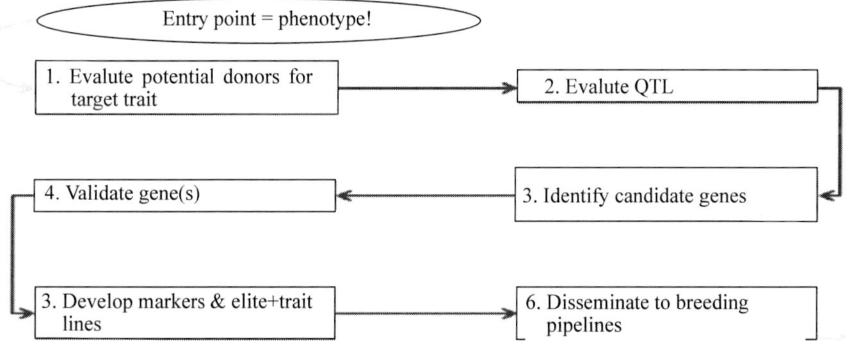

f) **Genetic Transformation:** This technique involves transfer of alien genes without involving sexual process. It is a rapid method of pre-breeding by which various plant traits can be easily improved through horizontal gene transfer.

g) **Cell Culture Techniques:** The somatic hybridization technique is useful in making new combinations of nuclear and cytoplasmic genes.

h) **Embryo rescue** and *in vitro* techniques.

i) **Involving bridge species**, which are crossable with both the parental species.

j) Application of exogenous plant growth regulators and immune-suppressors at post pollination stage.

k) Chromosome techniques such as translocations, alien additions lines (MAAL and DAAL), substitutions, homoeologous pairing and also irradiation to induce interchanges and gene transfer.

l) **Alloplasm development**: Transfer of cytoplasm (mt and cp genomes) from interspecific and intergeneric origin greatly enables the exploitation of cytoplasmic genetic male sterility system (CMS)- male fertility system. Contributions of cytoplasms of *Triticum timopheevi*, *Gossypium harknessii*, *Helianthus petiolaris, Oryza nivara* and *Cajanus scarabaeoides* in the development of CMS for harnessing heterosis are significant.

Examples

Genetic enhancement in wheat

Different genes for leaf and stem rust, stripe rust and powdery mildew resistance were transferred to hexaploid wheat from different *Triticum sp* and *Aegilops sp.*

Genes	Source
Sr 21, Sr 22	Triticum monococcum
Sr 2, Sr 9e, Sr 13	T. dicoccum
Sr 14, Sr 17, Sr 33	Ae. squarrosa

Germplasm enrichment of wheat by transfer of genes from Secale through translocations has enabled gene transfer for disease/insect resistance.

Translocations	**Resistance for**
1B. R - Tr	PM; stem, leaf and stripe
2BS. 2RL - Tr	Hessian fly
4BL. 5 RL - Tr	High copper tolerance
Irradiation-induced translocations	Hessian fly

Chromosome substitution method has helped for gene introgression in the following crops.

Crosses	Resistance
Triticum aestivum × Haynaldia	Powdery mildew
Trifolium repens × *T. nigrescens*	Cyst nematode
Beta vulgaris × Brassica. species.	Nematode
Cicer arietinum × *C. reticulatum*	Cyst nematode
Glycine max × *G. tomentella (homeologues)*	Cyst nematode

Genetic enhancement in Rice

Genes conferring resistance to insects and disease and other traits are introgressed as germplasm enhancement activity into cultivated rice through homoeologous pairing as shown below.

Species	Genes transferred
O. australiensis	BLB, BPH
O. latifolia	BLB, BPH, WBPH
O. brachyantha	BLB
O. officinalis	BPH, BLB, WBPH, GLH
O. minuta	Blast, BLB, BPH
O. perennis	CMS
O. nivara	Grassy stunt

Pre-breeding in cotton

- Development of Multiple Adversity Resistance which resistance to sucking pests, drought tolerance and improvement in fibre quality traits at Texas A & M University in U.S.A.
- The Australian cotton industry are facing a major challenge from Fusarium wilt attack in G. *hirsutum.* G. *sturtianum* is immune to the *Fusarium* pathogen – transfer of gene is difficult.
- The work of prebreeding has been initiated at CICR in 2002. Elite cultivars from all the three zones viz., North, Central and South have been taken up as base material and a crossing programme was initiated involving unadapted, unused G. *hirsutum* germplasm including wild species.
- It for Fibre quality traits viz., fibre length, fibre strength, micronaire value and Tolerance / resistance to sucking pests and bollworm complex.

Maize pre-breeding project

1. **Latin American Maize Project (LAMP):** LAMP is a real example of pre-breeding program, which includes 12 countries (Argentina, Bolivia,

Brazil, Colombia, Chile, U.S., Guatemala, Mexico, Paraguay, Peru, Uruguay and Venezuela). LAMP evaluated 15,000 accessions in the first stage, with close cooperation of the public and private sectors. Pioneer Hi-Bred International Company was decisive for the financial support of the project. The great genetic variability in Latin American maize is recognized, although there is much to be known, especially its potential and its significance for breeding approaches. Thus, maize breeders now have access to the most promising stocks identified by LAMP to expand the genetic base in maize.

2. **Hierarchical Open ended Population Enrichment (HOPE) System:** This system was first used by Kennenberg (1970) in maize. The main objectives of the HOPE system are: To provide a source of inbred lines that are genetically very different from those currently used in commercial breeding programmes, and the HOPE inbreds must be comparable in performance with current commercial inbreds both in *per se* performance and in hybrid combinations.

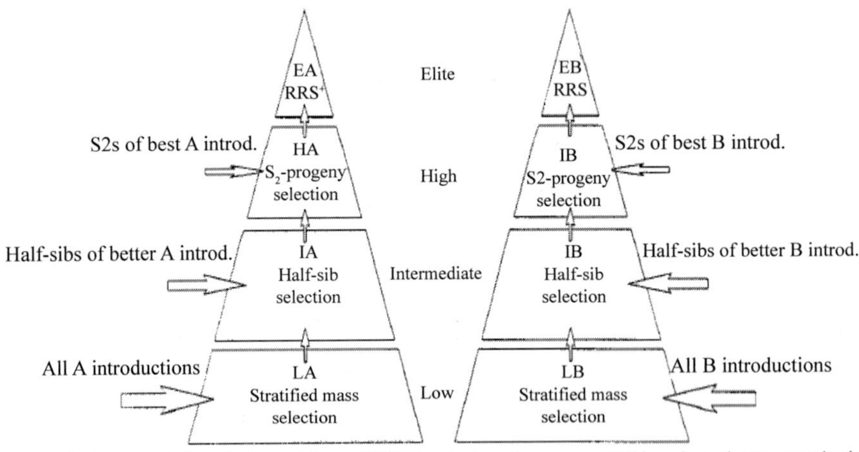

The hierarchical, open-ended population enrichment (HOPE) maize breeding system, RRS is reciprocal recurrent selection

Pre-breeding in barley

Recurrent Introgressive Population Enrichment (RIPE): RIPE was first adopted in 1990 by D.E. Falk in barley involving male sterile facilitated recurrent selection. The system consisted of one set of three hierarchical levels which like corn HOPE, was open-ended in that germplasm could move upward through the hierarchy and introductions could be added at the low level. However, the system was redesigned to intensify introgression at successive levels.

Some other crops

- **Sugarcane:** Inter-specific genetic stocks of S. *spontaneum, S. sinense and S. officinarum.*
- **Sorghum:** Incorporation of *Ethiopian, Sudanese land race traits into adapted Indian cultivars* has been done.
- **Potato** breeders developed day-long adapted populations which has the ability to long day tuberisation derived from Andcan tetraploid and diploid potato land races through mass selection. This is called Neo tuberosum.

Problems associated with Pre-breeding

- Cross incompatibility in inter-specific crosses.
- Stability barriers and chromosome pairing in hybrids have restricted the access to genes from wild species into cultivated ones.
- Linkage drag.
- Hybrid inviability and sterility.
- Small sample size of inter-specific hybrid population
- Restricted genetic recombination in the hybrid population.
- Lack of availability of donors for specific traits viz. resistance to diseases, pests and bollworm.
- Exchange and accessibility of cultivated species, germplasm material has become difficult due to legal restrictions like IPR.

Challenges and future prospects of pre-breeding

- The major challenges of pre-breeding are lack of characterization, evaluation of genetic diversity, documentation of data; inter species relationship and strong breeding program and funding sources.
- Genome mapping, decoding of genes and synteny among the genes could be assigned to conceal the stress tolerance and can be utilized for crop improvement.
- The potential of genetic transformation technique could be exploited to transfer the desired gene(s) form the tertiary gene pool and beyond.
- New breeding strategies and bioinformatics tools are required to use the information gathered from genetic and genome analysis programs for dealing with complex traits more effectively.

Comparison between Pre-breeding and traditional breeding

Pre breeding	Traditional Breeding
Pre-breeding is also known as genetic enhancement	Traditional breeding is also known as sustainable plant breeding
It leads to genetic enhancement of germplasm	It leads to development of productive cultivars/ hybrids
It leads to value addition	It does not leads to value addition
It leads to broadening the genetic base of the population	It leads to development of improved cultivars with narrow genetic base
The chief breeding method is backcross method	All breeding methods such as introduction, selection, hybridization and mutation are used.
The end products are improved germplasm line	The end product is cultivar or hybrid
The end product is used as parent for developing improved cultivar, hybrid	The end product is used for commercial cultivation
It involves adapted and non-adapted genotypes in crossing programme	It includes only adapted genotype
It is a long term breeding programme	It is a short or medium term breeding programme
It is taken up by public sector plant breeding organizations	It is taken up by both public and private sector organizations

References

Iqbal, A.M. et al. (2014). Pre-breeding and Population Improvement. An International Journal of Life Sciences. 2(03): 188-197.

Jain, S.K. and Omprakash (2019). Pre-breeding: A Bridge between Genetic Resources and Crop Improvement. Int.J.Curr.Microbiol.App.Sci. 8(02): 1998-2007. doi: https://doi.org/10.20546/ijcmas.2019.802.234.

Loknathan,T.R. et al. (2003). Genetic Enhancement In Cotton. Technical Bulletin from CICR, Bulletin No: 26.

5

Plant Genome Projects & Its Implications in Plant Genetic Resources

Abstract

The general aspect of genome projects revolves around to determine the complete genome sequence of an organism and to annotate protein-coding genes and other important genome-encoded features. Recombinant DNA and PCR technologies helped in preparation of molecular maps of many plant and animal genomes.

Keywords: *DNA sequencing few chromosomes, rapid life cycle*

Introduction

The objective of genomic research in any species is to sequence the whole genome and to decipher functions of all the different coding and non-coding seq,uences. The technology for large-scale DNA sequencing has enabled the scientists to undertake genome sequencing project in a realistic time scale. Since the time of first 'large' genome sequencing in bacteriophage λ in 1983, the projects on different groups have been completed. Eg. *Escherichia coli, Saccharomyces cerevisiae, Arabidopsis thaliana, Oryza sativa, Caenorhabditis elegans, Drosophila melanogaster, Mus musculus, Pan troglodytes, Homo sapiens.*

There are certain species which have been chosen for genome projects as model organisms. A model organism is a non-human species that is used to understand biological phenomena with a thought of gaining insight into the working of other organisms. They are chosen on the basis of their small genome in size, diploid genetics with a few chromosomes, rapid life cycle, easily transformed, well positioned in plant phylogeny, small stature for plant growth in a small space, large number of seeds produced, convenient for discovery of gene-trait discovery at low cost and speed and well investigated. They include,

- Monocots – *Oryza sativa*
- Dicots – *Arabidopsis thaliana*
- Temperate grasses – *Brachypodium*

- Pulses – *Medicago sativa*
- Biofuels – *Setaria italica*

Plant genome characteristics

Plant genomes like others, have almost the same number of genes. Their genome size shares a wide range, the carnivorous corkscrew plant, *Genlisea aurea*, at 63 Mb, the smallest known plant genome and the largest is that of the rare Japanese plant, *Paris japonica*, at 148,000 Mb. Plants tend to have roughly the same number of genes at about 32,000, despite this wide variation in genome size.

This broad range in genome size range appears to be driven by the proliferation of "copy-and-paste long terminal repeat (LTR) retrotransposons." Retrotransposons are DNA sequences that can copy themselves to RNA and then back into DNA. The copied DNA may then integrate back to the genome, increasing its size. Retrotransposons are found in people and other animals, but they are especially abundant in plants. The corn genome, for example, is bloated with 75% LTRs. Bladderwort's genome, on the other hand, is only 3% LTRs. Other reason for this great size range could be polyploidy.

Brachypodium is a close relative of other grasses, such as wheat, which are critical to world nutrition, but whose massive and complex genomes make them extremely hard to work with. *Brachypodium*, on the other hand, has one of the smallest known genomes among grasses, is easy to grow in the lab and manipulate genetically, and has a short life cycle. Thus, by working with *Brachypodium* instead, scientists can more quickly make advances that can then be used to improve vital cereal crops, such as wheat and oats.

Oilseed rape is the second most important vegetable oil in the world for cooking and industrial purpose. But it's an unusual hybrid that contains the entire genomes of two other plants: *Brassica rapa* and a closely related species called *Brassica oleracea*. By sequencing Chinese cabbage a variety of *Brassica rapa* (close relative of oilseed rape) access to half of oilseed rape's genes, is possible. All the *Brassica* relatives including broccoli, turnip, Brussels sprouts and cabbage are closely related, the insights scientists gain by from sequencing Chinese cabbage are expected to improve the breeding efficiency of a range of crops essential to global food security.

Arabidopsis Genome Project

Three American groups namely Meyerowitz, Somerville and Goodman at three different universities took the first step leading to the project on sequencing of the Arabidopsis genome later known as AGI. Initially the conventional 'clone-by- clone' was dominant approach but later shot gun approach was followed using cDNA libraries and ESTs. By 1999 the first report came on chromosome 2 and 4, but now the whole genome has been sequenced with all the information's.

1. The genome has approx. length of 145Mbp, of which genes are present (coding region is 2-2.5 kb) at each 4-5 kb interval.
2. Telomere and centromere regions are full of repeated DNA.
3. Centromeric region has transposons and pseudogenes.
4. Sometimes entire stretches of DNA and genes are duplicated between chromosomes.
5. Approx. 20% of the genes have signal sequences and target products into organelles such as chloroplasts or mitochondria. Entire mitochondrial genome of Arabidopsis is represented on chromosome no. 4.
6. It has about 20,000-50,000 genes of various functional groups.

Rice Genome Project

Scientists from Japan started the Rice Genome Programme (RGP) in 1991. In 1998 the second phase of RGP was launched, then 10 countries participated in the international Rice Genome Sequencing Programme (IRGSP).

- Monsanto (2000) produced the draft of rice genome of Japonica variety and was estimated to comprise of 420-466Mbp of DNA.
- In 2002 the draft sequence of Indica rice variety was published in Beijing Genomic Institute and Syngenta published a draft sequence of Japonica variety.
- IRGSP has made the progress in preparing contig maps employing YAC and BAC.
- From the sequence data available, it has been reported that the total number of genes in rice is not much larger than in Arabidopsis and most of the genes of two plants show homology.
- According to available data on rice from Gen Bank, 28,282,731 bases of sequences has been submitted from the rice EST sequencing projects.
- Average rice gene is 2.2 kbp containing 3.9 exons and 2.9 introns. The density of rice gene is one gene per 5.7 kbp.

1000 Plant Genomes Project

- Announced in 2008, shortly after the human 1000 Genomes Project, the **1000 Plant Genomes** was announced.
- **Project (1KP)** is a similar large-scale genomics using the high speed and efficiency of next-generation DNA sequencing.

- Headed by Dr. Gane Ka-Shu Wong and Dr. Michael, University of Alberta
- The project successfully sequenced the transcriptomes (expressed genes) of 1000 different plant species by 2014
- Goals of the Project
- As of 2002, the number of classified green plant species is around 370,000. Despite this number, very few of these species have detailed DNA sequence information.
- The 1000 Plant Genomes Project will produce a roughly a 100x increase in the number of species with available broad genome sequence.

Project Approach

- Using the 28 Illumina Genome Analyzer next-generation DNA sequencing machines at the Beijing Genomics Institute (BGI – Shenzhen, China).

Species selection

- The selection of plant species to be sequenced has nearly been compiled through an international collaboration of the various funding agencies and researcher groups expressing their interest in certain plants.
- There has been a focus on those plant species that are known to have useful biosynthetic capacity to facilitate the biotechnology goals of the project, selection of other species to fill in gaps and explain some unknown evolutionary relationships of the current plant phylogeny.
- In addition to industrial compound biosynthetic capacity, plant species known or suspected to produce medically active chemicals (such as poppies producing opiates) were assigned a high priority, to discover new pharmaceutical options.
- A large number of plant species with medicinal properties have been selected from traditional Chinese medicine (TCM).
- **Transcriptome vs. genome sequencing**
- Rather than sequencing the entire genome (all DNA sequence) of the various plant species, the project will sequence only those regions of the genome that produce a protein product (coding genes); the transcriptome.
- Justified by the focus on biochemical pathways where only the genes producing the involved proteins are required to understand the synthetic mechanism.

- Represent adequate sequence detail to construct very robust evolutionary relationships through sequence comparison.

Transcriptome sequencing

- mRNA (messenger RNA) is collected from a sample, converted to cDNA by a reverse transcriptase enzyme, and then fragmented so that it can be sequenced. Other than transcriptome shotgun sequencing, this technique has been called RNA-seq and whole transcriptome shotgun sequencing (WTSS) was used.
- Once the cDNA fragments are sequenced, they will be *de novo* assembled (without aligning to a reference genome sequence) back into the complete gene sequence by combining all of the fragments from that gene during the data analysis phase.

Funding

The project is funded by Alberta Innovates - Technology Futures, the Alberta Agricultural Research Institute (AARI), Genome Alberta, the University of Alberta, the Beijing Genomics Institute (BGI), and Musea Ventures (a USA-based private investment firm).

Potential limitations

- The project will not reveal information about gene regulatory sequence, non-coding RNAs, DNA repetitive elements, or other genomic features that are not part of the coding sequence, which may be the primary driver of trait differences seen between species.
- The amount of sequence representation for a given gene will be based on the expression level. This means that highly expressed genes get better coverage. The result, then, is that some important genes may not be reliably detected by the project if they are expressed at a low level yet still have important biochemical functions.
- Many plant species are known to have undergone large genome-wide changes through duplication of the whole genome, e.g. wheat genome. These duplicated genes may pose a problem for the *de novo* assembly of sequence fragments, because repeat sequences confuse the computer programs when trying to put the fragments together, and they can be difficult to track through evolution.

Table: List of crops under genome project

Name	**Scientific name**	**Year of sequenced**	**Genome size**	**No. of chromosomes**
Sugar beet	*Beta vulgaris*	2014	590MB	
Domesticated Tomato	*Solanum lycopersicum*	2012	900 MB	
Carnivorous bladderwort plant	*Utricularia gibba*	2013	82MB	
Wine Grape	*Vitis vinifera*	2007	500MB	19
Poplar	*Populus tricchocarpa*	2006	370MB	19
Sacred Lotus	*Nelumbo nucifera*		929 Mbp	
Tobacco	*Nicotiana benthamiana*		3GB	
Columbine	*Aquilegia sp.*		302 megabases	
Castor bean	*Ricinus communis*		320 megabases	10
Rubber tree	*Hevea brailiensis*	2013	2.15 GB	18
Melon	*Cucumis melo*	2012	450 MB	
Watermetlon	*Citrullus lanatus*	2012	353.5 MB	
Apple	*Malus domestica*	2010	742.3 MB	17
Pear	*Pyrus bretschneideri*	2012		17
Peach	*Prumus persica*	2010	227 MB	8
Chickpea	*Cicer arietinum*	2013	740 MB	
Soybean	*Glycine max*	2010	950 MB	20
Pigeon pea	*Cajanus cajan*	2011	833 MB	11
Rice	*Oryza sativa*	2002	370 MB	12
Barley	*Hordeum vulgare*	2012	5.1 GB	
Einkorn wheat	*Triticum urartu*	2013	5 GB	
Maize	*Zea mays*	2009	2.5 GB	10
Sorghum	*Sorghum bicolor*	2009	700 MB	10
Hemp	*Cannabis sativa*	2011	820 MB	
Musk melon	*Cucumis melo*	2012	450 MB	
Cucumber	*Cucumis sativus*	2009	367 MB	
Cassava	*Manihot esculenta*	2012	760 MB	
Common Bean	*Phaseolus vulgaris*	2013	520 MB	
Wild strawberry	*Gragaria vesca*	2011	240 MB	
Poplar	*Popuus trichocarpa*	2006	510 MB	
Potato	*Solanum tuberosum*	2011	844 MB	
Tobacco plant	*Nicotiana sylvestris*	2013	2.636 GB	

	Nicotiana tomentosformis	2013	2.682 GB	
Pepper	*Capsicum annuum*	2014	3.48 GB	
	Capsicum annuum var. glabriucculum	2014	3.48 GB	
Banana	*Musa acuminate*	2012	523 MB	
Wild banana	*Musa balbisiana*	2013	438 MB	
African oil palm	*Elaeis guineensis*	2013	1800 MB	
Date palm	*Phoenis dactylifera*	2011	658 MB	

Genome projects and its implications in PGR

- Phenotypic descriptors can be much clearly framed with an ample sum of genomic data
- Parentage and co- ancestry can be well defined
- DNA markers and sequencing produce quantitative views of genetic diversity and relationships more rapidly, comprehensively, and reproducibly than previous methods.
- They will create true gene banks or warehouses with functional inventories from largely uncharacterized gene pools.
- The traditional boundaries between gene pools have been challenged by genomic mapping.
- Identification of allele richness is made possible
- Technologies for large scale surveys of gene expression patterns, explore other dimensions of gene pools that have not been available for review.
- Instead of looking at individual genes, we will be compelled to review circuits of genes and their pathways to synthesize a meaningful comprehension of genetic variation.
- Epistasis may hinder the exploitation of desired trait from an exotic germplasm due to various reasons like environmental adaptation.
- A few bad genes for adaptation obscure or preclude the evaluation of many others.

6

Use of Genomic Tools in PGR Management

Abstract

Plant Genetic Resources (PGR) comprises of *Landraces, Farmers' varieties, Parental lines of hybrids Released varieties Wild and weedy species Primitive cultivars ,Genetic stocks Genetically modified organisms .*

Keywords: *Primitive cultivars, Genetic stocks, Genetically modified organisms*

Introduction

PGR management activities

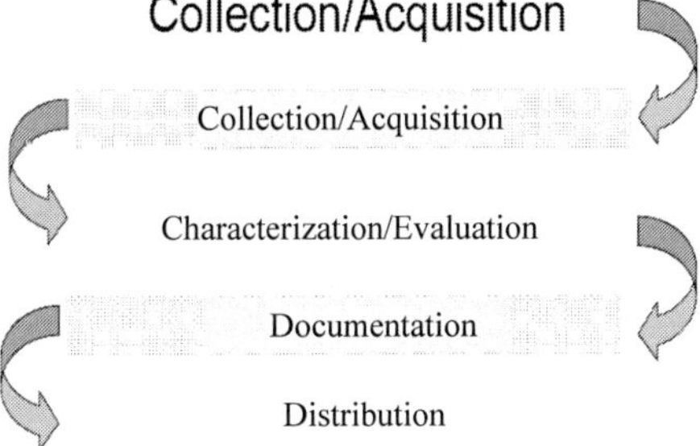

Genomics-driven approaches to Genebank management

- Pre-breeding can facilitate the characterization and utilization of PGRs.
- Allele mining involves searching useful alleles of genes from a wide germplasm collection.
- Transformation of germplasm collections into bio-digital resource centres like
- DNA Barcoding,
- DNA fingerprinting.

Pre-breeding

- Pre-breeding is the step before practical breeding.
- Aim - To introduce new desirable traits/genes into an adapted genetic background. This will broaden the genetic base in a breeding material in pace with environmental changes."
- The main link between the germplasm conservation and its use in plant breeding
- A vital contribution of pre-breeding is increasing the total genetic diversity in crops and finding specific genes and traits.
- Main link between conservation of PGR in gene bank collections and utilisation of these resources in agriculture and horticulture.
- Essential for linking genetic diversity arising from wild relatives and other unimproved materials for utilization.
- These activities are a collaboration between germplasm curator and the plant breeder who need to work germplasm collections and how new traits from these collections can bred into new varieties.

Main aims of Pre-Breeding

- Identify potentially useful genes in a well organized and documented genebanks.
- Design strategies that lead to development of an improved germplasm ready to use in varietal development.

Procedure of Pre-breeding and breeding approaches

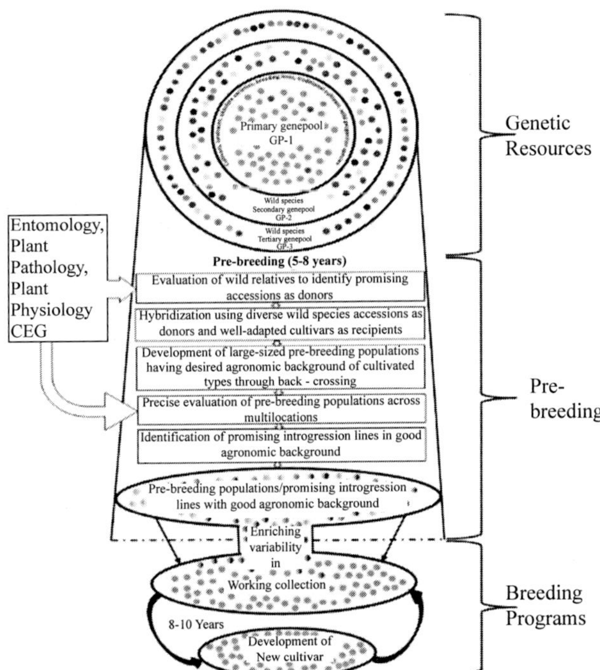

When Pre-breeding is becomes necessary ?

- When desired genes are available only in Genebank accessions those are not well-adapted to the target environment
- Closely related wild species that are easily crossed with the crop species
- More distant wild species which are more difficult to cross

When Pre- breeding is not needed?

- Commercially-available adapted and acceptable varieties
- Advanced selections, well-adapted to the target environment
- Genebank accessions that are well-adapted to the target environment

Applications of Pre-breeding

- Narrow genetic base results into the crop vulnerability to different biotic and abiotic stress. Pre breeding is adopted for broadening base, to reduce vulnerability.
- Wild species and crop wild relatives are the reservoir of the gene for cope with the changing climate, identification of this import a moving them from wild species into breeding populations when this appears to be the most effective strategy.

- Identification of novel genes in the unrelated species and transfer them using genetic transformation techniques.

Allele Mining

- Allele, it is a alternative forms of a gene found at a given locus on a chromosome. Mining is nothing but searching the new alleles in the wild germplasm.
- Searching of useful alleles of genes from a wide range of cultivars, related species and even across species.
- It is a research field aimed at identifying allelic variation of relevant traits within genetic resource collections. Kumari *et al.* (2018)

Need for Allele Mining?

- Progress in plant breeding in terms of development of **superior and high yielding varieties** of agricultural crops was made possible by accumulation of beneficial alleles from vast plant genetic resources existing worldwide.
- Still, a significant portion of these superior alleles were not utilized as these were left behind during evolution and domestication.
- This untapped genetic variation existing in wild relatives and land races of crop plants could be exploited gainfully for development of agronomically superior cultivars.
- Vast germplasm resources need to be relooked for novel alleles to further enhance the genetic potential of crop varieties for various agronomic traits.

True Allele Mining

- True allele mining is analysis of non coding and regulatory regions of the candidate genes in addition to analyzing sequence variations in the coding regions of genes

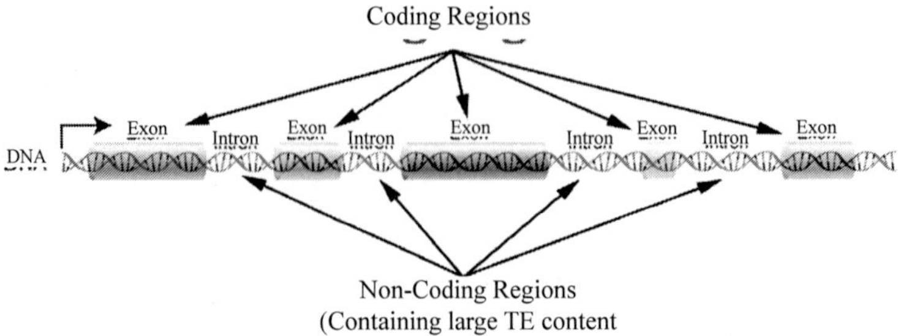

Importance of Allele Mining

- Tracing the evolution of alleles.
- Identification of new haplotypes and development of allele specific markers
- This capability will be important for giving breeders direct access to key alleles conferring:
  - Resistance to biotic stresses
  - Tolerance to abiotic stresses
  - Greater nutrient use efficiency
  - Enhanced yield
  - Improved quality

Steps Involved in Allele Mining

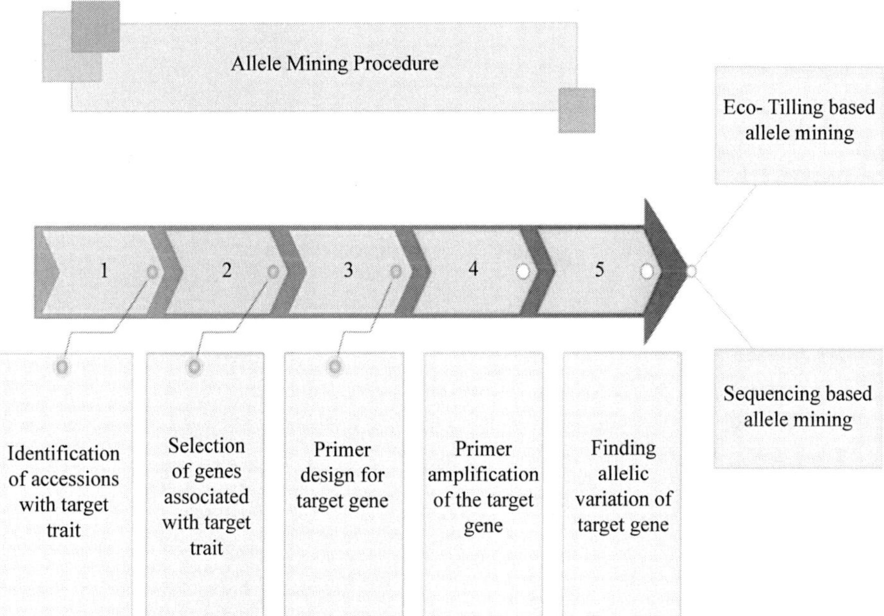

Approaches for Allele Mining

- Eco tilling based allele mining

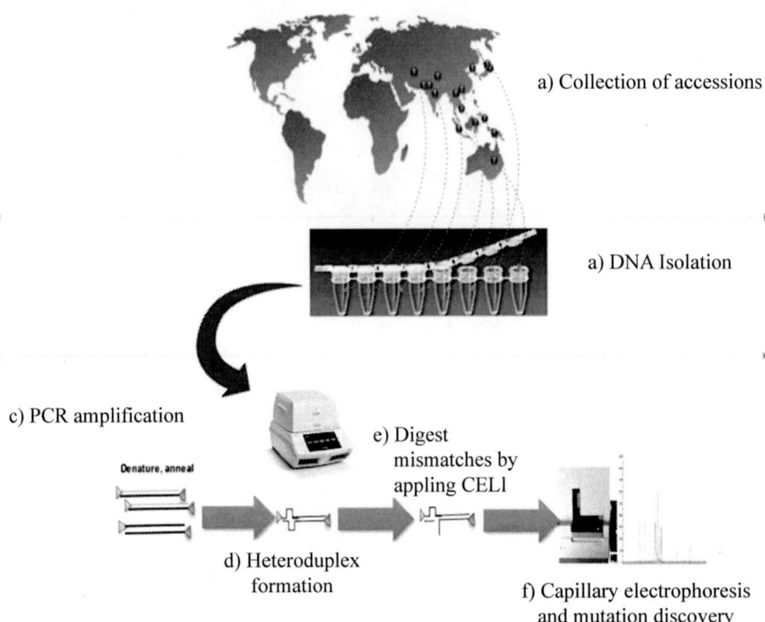

- Sequencing based allele mining

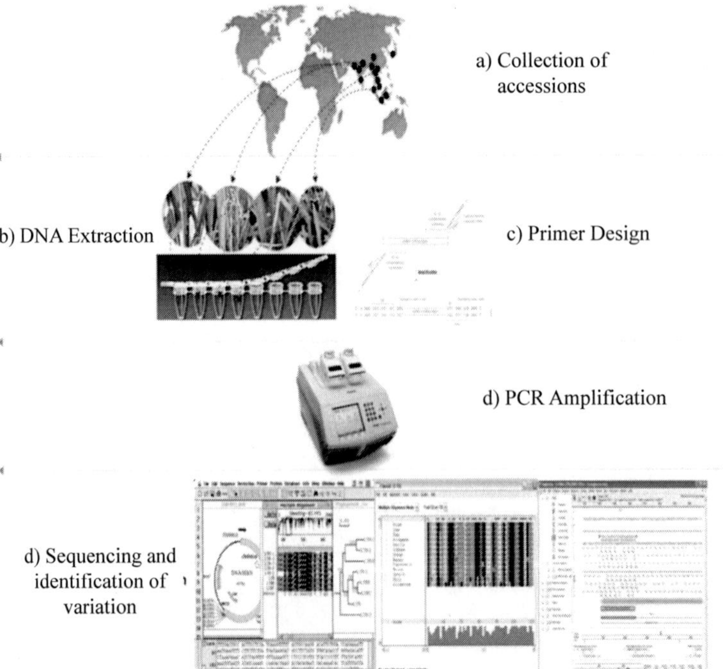

DNA barcoding

- Newer methods of Plant identification:
- Authentication at the DNA level - more reliability
- DNA is a stable macromolecule that is not affected by external factors
- The novel technique of identifying biological specimens using short DNA sequences (nuclear or organelle genomes) is called DNA barcoding.
- In plants, mitochondrial genes are not appropriate for DNA barcoding because they exhibit low mutation rates. In the chloroplast genome, the most promising being maturase K gene (matK) by itself or in association with other genes.
- Multi-locus markers such as ribosomal internal transcribed spacers (ITS DNA) along with matK, rbcL, trnH or other genes have also been used for species identification.

Procedure of DNA barcoding

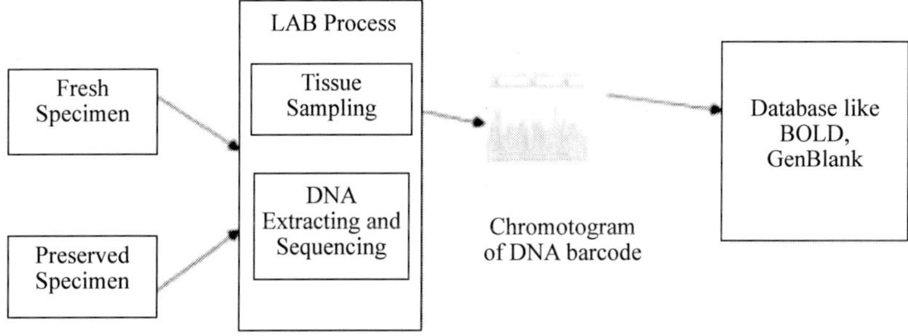

Freely accessible barcode libraries

- BOLD (The barcode of life data system) (http://www.barcodinglife.com)
- CBOL (Consortium for the barcode of life) (http://www.barcodeoflife.org/)
- iBOL (International Barcode of Life project) (http://www.ibol.org/)
- The GenBank online genetic sequence database (http://www.ncbi.nlm.nih.gov/genbank/)

References

Aubry, S. (2019). The Future of Digital Sequence Information for Plant Genetic Resources for Food and Agriculture. Frontiers in plant science, 10: 1046.

Gepts, P. (2006). Plant genetic resources conservation and utilization. Crop Science, 46(5): 2278-2292.

Mascher, M., Schreiber, M., Scholz, U., Graner, A., Reif, J. C., & Stein, N. (2019). Genebank genomics bridges the gap between the conservation of crop diversity and plant breeding. Nature genetics, 51(7): 1076-1081.

Vijayan, K., & Tsou, C. H. (2010). DNA barcoding in plants: taxonomy in a new perspective. Current Science, 1530-1541.

Van Treuren, R., & van Hintum, T. J. (2014). Next-generation genebanking: plant genetic resources management and utilization in the sequencing era. Plant Genetic Resources, 12(3): 298-307.

7

Need for Novel Genes for Genetic Enhancement of Crop Species

Abstract

Genetic enhancement: Transferring useful genes from exotic or wild types into agronomically acceptable background coined by **Jones** *(1983).vRick (1984) used the term pre-breeding or developmental breeding to describe the same activity.*

Keyword*: Pre-breeding, Genetic enhancement, Genepool*

Introduction

- Thus "genetic enhancement" or "pre-breeding" refers to the transfer or introgression of genes or gene combinations from unadapted sources into breeding materials (FAO, 1996).
- Genetic enhancement plays an important role in utilizing unadapted and unutilized germplasm collections and creating vast genetic variability for development of productive cultivars / hybrids.
- Objective of Genetic Enhancement is broadening the genetic base and increases genetic diversity
- Gene pool – Harlan and de Wet (1971).
- Primary gene pool (GP-1): Members of this gene pool are probably in the same "species" (in conventional biological usage) and can intermate freely.
- Secondary gene pool (GP-2): Members of this pool are probably normally classified as different species than the crop species under consideration (the primary gene pool).
- Tertiary gene pool (GP-3): Members of this gene pool are more distantly related to the members of the primary gene pool. The primary and tertiary gene pools can be intermated, but gene transfer between them is impossible without the use of "rather extreme or radical measures such as embryo rescue (or embryo culture, a form of plant organ culture) induced polyploidy (chromosome doubling) bridging crosses (e.g., with members of the secondary gene pool).

Rice

S. No	Gene	Function	Source	References
1.	*PSTOL1*	Phosphorus starvation tolerance	*Oryza rufipogan*	Neelam et al., 2017
2.	*MSR2*	Drought and salt tolerance	*Oryza rufipogan*	Sanchez et al., 2013
3.	*Bph10* and *Bph18*	Brown plant hopper	*O. australzensis*	Jena et al., 2006
4.	*Bph13*	Brown plant hopper	*O. officinalis*	Jena et al., 2006
5.	*Bph14*	Brown plant hopper	*O. officinalis*	Ishii et al., 1994
6.	*Pi40*	Fungus resistance	*O. australzensis*	Jeung et al., 2007
7.	*Pi9*	Blast	*Oryza minuta*	Amante-Bordeos et al., 1992
8.	*Xa21, Xa27, Xa32,Xa35*	Bacterial blight	*Oryza barthii*	Gu et al., 2004
9.	*Xa23*	Bacterial blight	*O. rufipogon*	Zhang et al.,2005
10.	*Xa30*	Bacterial blight	*O. rufipogon*	Jin et al., 2007
11.	*Xa29* `	Bacterial blight	*O. officinalis*	Tan et al., 2004
12.	*Xa32*	Bacterial blight	*O. Ustraliensis*	Zheng et al., 2009
13.	*Xa36*	Bacterial blight	*O. barthii*	Miao et al., 2010
14.	*yld1.1 and yld2.1*	Yield enhancing QTLs	*O. rufipogon*	Xiao et al., 1998
15.	*gpp1.1 + sd 1 locus*	Yield	O. rufipogon	Cho et al., 2003
16.	*SUB1*	Submergence	*O. rufipogon and O. nivara*	Niroula et al., 2012
WHEAT				
17.	*TmHKT1*	Salt tolerance	*Triticum monococcum*	James, Davenport, & unns, 2006

18.	*HSP*	Heat tolerance	T. monococcum	Vierling and Nguyen 1992
19.	*Pm1*	Powdery mildew Resistance	*T. monococcum*	Yao et al., 2007
20.				
21.	*Rp1td*	Rust resistance	*Z. diploperennis*	Bergquist, 1981
22.	*Ht3*	Northern corn leaf blight resistance	*Z. diploperennis*	Hooker, 1981
23.	*wip1, RP1 and chitinase*	Insect tolerance (FAW)	*Z. mays spp. Parviglumis*	Szczepaniec et al., 2013
24.	*bin 4.07 (QTL)*	Sogray leaf spot resistance	*Z. diploperennis*	Chavan and Smith, 2014
SOYBEAN				
25.	*rhg1-a, rhg1-b, and Rhg4 (cqSCN-006 & cqSCN-007)*	Soybean cyst nematode tolerance	*G. soja*	Kim et al., 2011
26.	*Rpp4* *C4 & R gene*	Soybean rust	*G. tomentella,* *G. tabacina*	Meyer et al., 2009
27.	*R gene*	Sclerotina Stem Rot resistance	*G. tabacina* *G. tomentella*	Hartman et al., 2000
28.	*GsWRKY20* *GsCHX1*	Salt tolerance	*G. Soja*	Tang et al., 2014
29.	*GsJAZ2*	Salt and alkali stress	G. soja	Zhu et al., 2012
30.				
31.	*H6, H1 & MU 8b*	Tolerance to Jassids	*G. raimondii*	Saunders, 1965 Zafar et al., 2009
32.	*Resistance genes in Chr-11 and Chr 14*	Root-knot nematode resistance Reniform nematode resistance	*G. longicalyx* *G. stocksii*	Meyer et al., 2009
33.	*B6*	Bacterial blight resistance	*G. arboretum*	Zafar et al., 2009
34.	*R gene*	Rust resistance	*G. anomalum*	Blank and Leathers, 1963
35.	*WRKY genes*	Salt tolerance	*G. aridum* *G. davidsonii* *G. tomentosum*	Fan et al., (2015)

TOMATO				
36.	*Ty genes (1 to 5)*	Tomato yellow leaf curl virus	*Solanum chilense* *S. pimpinellifolium* *S. peruvianum*	Zamir et al., 1994 Black, & Hanson, 2002 Warmink, & Hille, 2005
Biotic and abiotic stress tolerance genes from Crop Wild Relatives				
37.	*IRIP* *C- repeat binding factor (CBF3) genes*	Cold tolerance	*Brachypodium Distachyon*	Li et al., 2012
38.	*sucrose synthase gene, glucose-1-Phosphate adenylyltransferase Gene*	Drought tolerance	*B. distachyon*	Verelst et al., 2013
39.	*LcDREB3a transcription factor gene*	Salt tolerance	*Leymus chinensis*	Xianjun et al., 2011
40.	*VpWRKY3*	Biotic and abiotic stress	*Vitis pseudoreticulata*	Zhu et al., 2012
41.	*Lr21*	leaf rust resistance	*Aegilops tauschii*	Ling et al., 2004
42.	*6-SFT*	Drought and cold Tolerance	*Agropyron cristatum*	Shan and Liang 2010;
43.	Dhn1 and Dhn6	Drought tolerance	*Hordeum. Spontaneum*	Suprunova et al. 2004
44.	*Hsdr4*	Drought tolerance	*H. Spontaneum*	Suprunova et al. 2007
45.	*FhbLoP*	Fusarium head blight resistance	*Thinopyrum elongatum*	Chen et al., 2013
46.	*Serine/threonine kinase gene Stpk-V*	Powdery mildew Resistance	*Haynaldia villosa*	Cao et al., 2011
47.	*Pc39, Pc45 and Pc94 genes*	Crown rust resistance	*Avena barbarata*	Cabral and Park, 2014
Genes from other sources				
48.	*aad-1*	2,4-D herbicide tolerance	*Delftia acidovorans*	

49.	*pg (polygalacturonase)*	Delayed fruit softening	*Lycopersicon esculentum*	ISAAA
50.	*sam-k (S-adenosylmethionine hydrolase)*	Delayed ripening/ senescence	*Escherichia coli bacteriophage T3*	
51.	*dmo (dicamba mono-oxygenase)*	Dicamba herbicide tolerance	*Stenotrophomonas maltophilia*	
52.	*cspB (cold shock protein B)*	Drought stress tolerance	*Bacillus subtilis*	
53.	*Hahb-4*	Drought stress tolerance	*Helianthus annuus*	
54.	*psy1*	Enhanced Provitamin A Content	*Zea mays*	
55.	*mcry51Aa2*	Hemipteran Insect Resistance	*Bacillus thuringiensis*	
56.	*OsGA2ox1*	Inhibited stem elongation	*Oryza sativa L. cv Nipponbare*	
57.	*RB*	Late blight disease resistance	*Solanum bulbocastanum*	
58.	*cry1A, cry1Ab, cry1Ac, cry1F, cry2Ab2, cry9C and vip3A(a)* *pinII (protease inhibitor)*	Lepidopteran insect resistance	*Bacillus thuringiensis* *Solanum tuberosum*	
59.	*barnase* *Dam zm-aa1*	Male sterility	*Bacillus amyloliquefaciens* *E. coli* *Zea mays*	
60.	*cordapA*	Modified amino acid (lysine)	*Corynebacterium glutamicum*	
61.	*fatb1-A, gm-fad2-1*	Modified oil/fatty acid	*Glycine max*	
62.	*gbss (granule-bound starch synthase)*	Modified starch/ carbohydrate	Solanum tuberosum	

Advantages of Genetic Enhancement

- It also leads to creation of new variability.
- Broadening the genetic base, to reduce vulnerability.
- Introgression genes from wild species into breeding populations.
- Creation of polyploidy.
- Identification and transfer of novel genes from unrelated species using genetic transformation techniques.